Engineering Materials

This series provides topical information on innovative, structural and functional materials and composites with applications in optical, electrical, mechanical, civil, aeronautical, medical, bio- and nano-engineering. The individual volumes are complete, comprehensive monographs covering the structure, properties, manufacturing process and applications of these materials. This multidisciplinary series is devoted to professionals, students and all those interested in the latest developments in the Materials Science field.

More information about this series at http://www.springer.com/series/4288

Onoyivwe Monday Ama · Suprakas Sinha Ray
Editors

Nanostructured Metal-Oxide Electrode Materials for Water Purification

Fabrication, Electrochemistry and Applications

 Springer

Editors
Onoyivwe Monday Ama
Department of Chemical Sciences
University of Johannesburg
Johannesburg, South Africa

Suprakas Sinha Ray🆔
Department of Chemical Sciences
University of Johannesburg
Johannesburg, South Africa

Centre for Nanostructures and Advanced
Materials, DSI-CSIR Nanotechnology
Innovation Centre, Council for Scientific
and Industrial Research
Pretoria, South Africa

ISSN 1612-1317   ISSN 1868-1212   (electronic)
Engineering Materials
ISBN 978-3-030-43348-2   ISBN 978-3-030-43346-8   (eBook)
https://doi.org/10.1007/978-3-030-43346-8

This Springer imprint is published by the registered company Springer Nature Switzerland AG
The registered company address is: Gewerbestrasse 11, 6330 Cham, Switzerland

Preface

Over the last two decades, "nanostructured materials" or "nanostructures" have become popular terms, not only among researchers, but also the general public. Owing to their extraordinary and unexpected behavior; nanostructures have gained tremendous attention in various fields such as water purification, sensors, electronics, health care, and biomedical applications. In this direction, there is a recent focus on the modification of nanostructured properties of some solvent electrolyte combinations that are commonly used in electrochemical characterization. However, the electrolyte provides the pathway for ions to flow between and among electrodes in the cell to maintain charge balance for the degradation of organic dyes with the use of metal oxide. Since water treatment is becoming more challenging owing to the complexity of the pollutants present within. This necessitates the need to seek alternative and complementary approaches to the removal of these organic dyes. This book will contribute to this need in our community, which cuts across electrochemistry, water science, materials science, and nanotechnology. It successfully highlights the applicability of the fabricated electrode and nanocomposites in electrochemical technology.

This book is structured in ten chapters. Chapter 1 focuses on the dynamic degradation efficiency of major organic pollutants such as pharmaceutical wastes and synthetic dyes from wastewater. Chapter 2 describes the synthesis and fabrication of photoactive nanocomposite electrodes for the degradation of wastewater pollutants. Chapter 3 is dedicated to the essence of electrochemical measurements on corrosion characterization and electrochemistry applications. The particular focus is on photoelectrochemical technique. Different kinds of electrochemical cells and their definitions of specific physical quantities are presented in Chap. 4. The synthesis and properties of various metal oxide nanoparticles in electrochemical applications are summarized in Chap. 5. Chapter 6 focuses on the construction of different biosensors using metal oxide nanoparticles for the detection of various harmful biological molecules in water. In Chap. 7, the authors report multiple kinds of metal oxide nanomaterials for electrochemical detection of heavy metals in water. Chapter 8 provides various applications of metal oxide electrodes and Chap. 9 focuses applications of modified metal oxide electrodes in photoelectrochemical removal of organic pollutants from wastewater. The final section, Chap. 10, provides metal oxide nanocomposites for adsorption and photochemical degradation of pharmaceutical pollutants.

In summary, this book addresses fundamental issues that persist concerning nanostructured metal oxide-based electrode design which cuts across electrochemistry, water science, materials science, and nanotechnology. This book is ideal for water scientists, material scientists, researchers, engineers (chemical and civil) including under- and post-graduate students who are interested in this exciting field of research. Moreover, this book will also help industrial researchers and R&D managers who want to bring advanced nanostructured metal oxide-based electrodes into the market.

Finally, we express our sincerest appreciation to all authors for their valuable contribution, as well as reviewers for their critical evaluation of proposals and chapter manuscripts. Our special thanks go to Dr. Zachary Evenson, Associate Editor at Springer Nature for his suggestions, guidance, and advice during various stages of book preparation, organization, and production of this book. The financial support from the Council for Scientific and Industrial Research, the Department of Science and Innovation, and the University of Johannesburg is highly appreciated.

Johannesburg, South Africa   Onoyivwe Monday Ama
Johannesburg, South Africa   Suprakas Sinha Ray
Pretoria, South Africa

Contents

Editors and Contributors

About the Editors

Dr. Onoyivwe Monday Ama received a Ph.D. degree in Chemistry at the University of Johannesburg, South Africa 2017. He is the author of eighteen international articles in high impact international journals. Dr. Ama multi-disciplinary work cuts across electrochemistry, water science, materials science, and nanotechnology. It successfully highlights the applicability of exfoliated graphite nanocomposites in electrochemical technology. His contribution paves the way for the development of electrochemical reactors, powered by solar light in our society, for water treatment shortly.

Prof. Suprakas Sinha Ray is a chief researcher in nanostructured materials at the Council for Scientific and Industrial Research (CSIR) with a Ph.D. in physical chemistry from the University of Calcutta in 2001 and Manager of the Centre for Nanostructures and Advanced Materials, DSI-CSIR Nanotechnology Innovation Centre. Ray's current research focuses on advanced nanostructured materials and their applications. He is one of the most active and highly cited authors in the field of polymer nanocomposite materials, and he has recently, been rated by Thomson Reuters as being one of the Top 1% most impactful and influential scientists and Top 50 high impact chemists (morethan two million chemistry worldwide).

Prof. Ray is the author of 5 books, co-author of 3 edited books, 32 book chapters on various aspects of polymer-based nanostructured materials & their applications, and author and co-author of 370 articles in high-impact international journals, 30 articles in national and international conference proceedings. He also has 6 patents and 7 new demonstrated technologies shared with colleagues, collaborators, and industrial partners. So far, his team commercialized 19 different products. His honors and awards include South Africa's most *Prestigious 2016 National Science and Technology Award* (NSTF); **Prestigious 2014 CSIR-wide Leadership** award; **Prestigious 2014 CSIR Human Capital development award**; *Prestigious 2013 Morand Lambla Awardee* (top award in the field of polymer processing worldwide), International Polymer Processing Society, USA. He is also appointed as Extraordinary Professor, University of Pretoria and Distinguished Visiting Professor of Chemistry, University of Johannesburg.

Contributors

Feyisayo V. Adams Department of Petroleum Chemistry, American University of Nigeria, Yola, Nigeria;
Department of Metallurgy, School of Mining, Metallurgy and Chemical Engineering, Faculty of Engineering and the Built Environment, University of Johannesburg, Doornfontein, South Africa

Uyiosa Osagie Aigbe Department of Physics, College of Science, Engineering and Technology, University of South Africa, Pretoria, South Africa

Seyi Philemon Akanji Department of Chemical Science, University of Johannesburg, Johannesburg, South Africa

Onoyivwe Monday Ama Department of Chemical Science, University of Johannesburg, Johannesburg, South Africa;
Department of Chemical Engineering, Vaal University of Technology, Vanderbijlpark, South Africa;
DST-CSIR National Center for Nanostructured Materials, Council for Scientific and Industrial Research, Pretoria, South Africa

William Wilson Anku CSIR-Water Research Institute, Accra, Ghana

V. Chauke Polymers and Composites, Materials Science and Manufacturing, Council for Scientific and Industrial Research, Pretoria, South Africa

A. Chetty Polymers and Composites, Materials Science and Manufacturing, Council for Scientific and Industrial Research, Pretoria, South Africa

L. Chimuka Molecular Sciences Institute, School of Chemistry, University of the Witwatersrand, Johannesburg, South Africa

Chikaodili Chukwuneke Department of Petroleum Chemistry, American University of Nigeria, Yola, Nigeria

David Jacobus Delport Department of Chemical, Metallurgical and Materials Engineering, Tshwane University of Technology, Pretoria, South Africa

Ikenna Chibuzor Emeji Department of Chemical Engineering, Vaal University of Technology, Vanderbijlpark, South Africa;
Department of Chemical Science, University of Johannesburg, Johannesburg, South Africa

Azeez Olayiwola Idris Nanotechnology and Water Sustainability Research Unit, College of Science, Engineering and Technology, University of South Africa, Johannesburg, South Africa

Oluwagbenga T. Johnson Department of Metallurgy, School of Mining, Metallurgy and Chemical Engineering, Faculty of Engineering and the Built Environment, University of Johannesburg, Doornfontein, South Africa;
Department of Mining and Metallurgical Engineering, University of Namibia, Windhoek, Namibia

Khotso Khoele Department of Chemical, Metallurgy and Material Engineering, Tshwane University of Technology, Pretoria, South Africa;
Department of Chemical Science, University of Johannesburg, Johannesburg, South Africa;
Department of Chemical, Metallurgical and Materials Engineering, Tshwane University of Technology, Pretoria, South Africa

Joshua O. Madu Department of Petroleum Chemistry, American University of Nigeria, Yola, Nigeria

L. Mdlalose Polymers and Composites, Materials Science and Manufacturing, Council for Scientific and Industrial Research, Pretoria, South Africa

P. Msomi Department of Applied Chemistry, University of Johannesburg, Johannesburg, South Africa

N. Nomadolo Polymers and Composites, Materials Science and Manufacturing, Council for Scientific and Industrial Research, Pretoria, South Africa

Peter Ogbemudia Osifo Department of Chemical Engineering, Vaal University of Technology, Vanderbijlpark, South Africa

Suprakas Sinha Ray DST-CSIR National Center for Nanostructured Materials, Council for Scientific and Industrial Research, Pretoria, South Africa;
Department of Chemical Science, University of Johannesburg, Johannesburg, South Africa

K. Setshedi Polymers and Composites, Materials Science and Manufacturing, Council for Scientific and Industrial Research, Pretoria, South Africa

Chapter 1
Dynamic Degradation Efficiency of Major Organic Pollutants from Wastewater

Khotso Khoele, Onoyivwe Monday Ama, Ikenna Chibuzor Emeji, William Wilson Anku, Suprakas Sinha Ray, David Jacobus Delport, and Peter Ogbemudia Osifo

Abstract Due to the impact of photo-active materials on removal of disposed pharmaceutical waste and synthetic dyes from wastewater, attention of numerous researchers has been hugely turned onto dynamic degradation efficiency of various photocatalysts constituents. On the application of photo-active composites, they are predominantly synthesized from nanoparticles, and fabricated to photo-active semiconductors. At the center of overcoming limitations of photocatalysis technique which encompass rapid photo generation electron-hole pairs, high operational costs and utilization of small portion of solar energy, synergism of photocatalysist technique to electrochemical oxidation became a necessity for a better degradation efficiency. Hence, this chapter addresses a dynamic degradation efficiency provided by combined photocatalysists within variety of synthetic dyes and pharmaceutical products and application of photoelectrochemical oxidation technique. An emphatically sensitized discussion is on variety of synthetic dyes, pharmaceutical products, environmental concerns and photocatalysists. Overall, a degradation efficiency based on paramount factors has thoroughly been clarified.

K. Khoele · O. M. Ama (✉) · I. C. Emeji · S. S. Ray
Department of Chemical Science, University of Johannesburg,
Doornfontein, 2028, Johannesburg, South Africa
e-mail: onoyivwe4real@gmail.com

K. Khoele · D. J. Delport
Department of Chemical, Metallurgical and Materials Engineering,
Tshwane University of Technology, Pretoria, South Africa

O. M. Ama · S. S. Ray
DST-CSIR National Center for Nanostructured Materials,
Council for Scientific and Industrial Research, Pretoria 0001, South Africa

O. M. Ama · I. C. Emeji · P. O. Osifo
Department of Chemical Engineering, Vaal University of Technology,
Private Mail Bag X021, Vanderbijlpark 1900, South Africa

W. W. Anku
CSIR-Water Research Institute, P.O. Box M.32, Accra, Ghana

© Springer Nature Switzerland AG 2020
O. M. Ama and S. S. Ray (eds.), *Nanostructured Metal-Oxide Electrode Materials for Water Purification*, Engineering Materials,
https://doi.org/10.1007/978-3-030-43346-8_1

1.1 Introduction

Major organic pollutants of the wastewater are synthetic dyes and pharmaceutical waste. Dyes are synthetically produced for textiles manufacturing, paper printing, leathers applications, ceramics' making, cosmetics, inks and food processing [1–10]. From eventual usage, about 15% of organic dyes is lost annually into the wastewater [11–15]. On the other hand, pharmaceutical products which are produced intendedly for health related issues [16–20] also end up being into the wastewater. This comes from a dispose of pharmaceutical waste mainly from hospitals, house-holds and industries.

Complications emanating from wastewater comprised of organic dyes and pharmaceutical waste consists of hazardous impact on human and animal health [21–28]. Furthermore, organic dyes and pharmaceutical waste also have negative implications on the environment [29, 30]. These negativities necessitate a use of photo-active metal oxides to remove pollutants from the wastewater.

Photocatalysts are metal oxides which absorb light under solar application, and are often called photocatalyst semiconductors. Photocatalysts semiconductors encompass suitable bandgaps and highly photodegradation efficiency for diverse wastewater pollutants [31]. Some of the mostly utilized photocatalyst semiconductors include, but not limited to: TiO_2, MnO_2, WO_3, ZnO, Fe_2O_3, NiO, $SrZrO_3$, $BiOI/Ag_3VO_4$, $Ag_2O/Ag_3VO_4/AgVO_3$, etc. [32–40]. An application of photocatalyst semiconductors on wastewater treatment is carried by various methods, and photocatalysis technique is one of the major ones.

Photocatalysis is an undisputed green-making wastewater treatment technique which has massively been used by numerous researchers due to its significance. Earlier, interest of engaging photocatalysists semiconductors under photocatalysis technique was mostly on dye decolourization. Wang et al. [41] studied number of different dyes to characterize photocatalytic degradation process particularly on dyes' decolourization by photocatalysis. On utilization of Degassa P25 TiO_2 as semiconductor photocatalyst, effective removal of colors from all dyes was noted. Zainal et al. [42] investigated photocatalytic degradation of Novel guar gum/Al_2O_3 in degradation of malachite green (MG) dye using photocatalysis technique, and it was obtained that the guar gum/Al_2O_3 significantly degrade MG between 80 and 90%. Wang et al. [43] synthesized green SnO_2 nanoparticle and examine it for a degradation ability of the predominantly utilized dye methylene blue (MB). From engaged photocatalysis measurements, analysis revealed that the fabricated SnO_2 nanoparticles effectively remove MB from the wastewater. Furthermore, Han et al. [44] utilized nonconventional SnO_2 nanoparticles on degradation of MB using photocatalysis, and about 90% degradation efficiency was yielded. Choi et al. [45] also utilized TiO_2 on degradation of wastewater, and it was discovered that TiO_2 effectively remove organic pollutants. On the other hand, Bessegato et al. [46] distinguished effectiveness of dyes' removal between nano-$SrTiO_3$ and nano-TiO_2. From the analysis, nano-TiO_2 proves overall superiority. Nonetheless, as significant as photocatalysis technique is, it possesses negative propensities. First, a recombination rate of electron-hole pairs occurs, and that affects effectiveness of Photoanodes. Second, it has inferior solar light absorption, and that consequent to low efficiency [47]. Hence, the photocatalysis technique is incorporated to the electrochemical oxidation technique.

Coupling photocatalysis technique to electrochemical oxidation gives a synergistic photoelectrochemical technique. On utilization of this method, an immobilization of the photocatalyst semiconductor on a conducting material occurs from the application of a bias potential. Photo electrons from photocatalyst semiconductor react with released oxygen to develop oxygen reactive radicals which react with protons to generate H_2O^-. Consequently, the radicals undergo chemical reactions with the wastewater pollutants [48]. The whole process is accompanied by the generation of hydroxyl species from the synergy of two techniques, and that significantly improves process efficiency and effectiveness of engaged photoanode [49]. Hence, photoelectrochemical technique is presently reckoned as a chief technique on wastewater treatment.

On the continual wastewater treatment using photoelectrochemical technique, degradation efficiency using modified photoelectrodes within variety of dyes and pharmaceutical products is the core research area presently. Hence, this chapter has an emphatic discussion on variety of organic dyes, pharmaceutical products, environmental concerns, photocatalysist semiconductors. Overall, a degradation efficiency based on different dyes and methods has thoroughly been clarified.

1.2   Dyes

In ancient times, natural dyes were utilized mostly for food processing, and there was narrow usage of synthetic dyes. Natural dyes generally contain organic and inorganic materials which are produced from animal and plant products, and they are less harmful on their dispose to the environment. As the world's population grows, however, the use of the synthetic dyes inevitably expands.

1.2.1   Synthetic Dyes

The major industrial application of synthetic dyes is on textiles, leather, fabrics, paper, cosmetic, electroplating, distillation, pharmaceutical products, etc. [50–60]. Synthetic dyes are classified into about ten parts: acid, basic, direct, disperse, metallic, mordant, pigment, reactive, solvent, Sulphur and vat dyes. Nonetheless, acid and basic dyes are more popular due to their high-fastness, brilliance and versatility of application [61]. Thus, most research work is done on both acidic and basic synthetic dyes.

1.2.2 Some Predominantly Used Acidic Dyes

- Acidic Red 114
- Acidic Orange 7
- Acidic Orange 8
- Acidic Red 27
- Acidic Orange 6
- Acidic Blue 1.

1.2.3 Basic Dyes

- Methylene Blue
- Methylene Orange
- Basic Red
- Basic Violet
- Basic Blue 41.

1.3 Pharmaceutical Products

Pharmaceutical products (PPs) are significantly produced to ease both human and animal health. However, their eventual usage, storage and disposal lead to various problems. Minimally, the disposal of PPs eventually transports them into the wastewater. Figure 1.1 summarizes sources that lead PPs end up being into the wastewater. Once this occurs, an opposite of what they were originally produced for becomes a reality. People, animal and plants exposed to excreted PPs could face a definitive health threads. Furthermore, excretion of PPs is also an environmental hazard [63–66]. Hence, a complete abatement of PPs from the wastewater is paramount.

1.3.1 Ways Pharmaceutical Products Go into Wastewater

PPs present into wastewater from three major sources, and there are hospitals, households and drug factories.

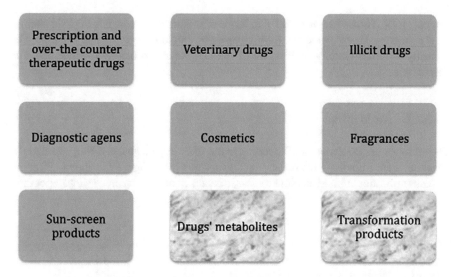

Fig. 1.1 Sources that lead pharmaceutical products to be present in the wastewater [62]

1.3.1.1 Households

When PPs are used in households for medication, facial products, etc. they do not fully become metabolized [67]. This causes them to be excreted from the body, and that consequent to them being into the wastewater. Furthermore, the dumping of use PPs and disposal of the expired medication lead to PPs being into the wastewater.

1.3.1.2 Hospitals

Hospitals encounter occurrences of the expiring PPs on their shelves, and mandatorily dispose them through hospital drainages.

1.3.1.3 Drug Factories

Manufacturing of PPs involve discharging point, and that eventually lead to PPs into the wastewater.

1.3.2 Types of Pharmaceutical Products Polluting Wastewater

There are about four PPs which are generally found in wastewater, and these are: antibiotics, steroidal hormones, blood-lipid regulators and anti-inflammatory

agents [68]. The nature of each PP in terms of intended use consists of toxic chemical elements. These include sodium, potassium, calcium, chloride and bromide. On
mostly reported toxicity of PPs from the wastewater, the mostly touched PPs are
antibiotics.

1.3.2.1 Antibiotics

Antibiotics are compounds released from the microorganisms, and they are mainly
used to treat infections from their various usages [69]. Figure 1.2 shows different
areas antibiotics applied at. Nonetheless, toxicity of antibiotics in wastewater comes

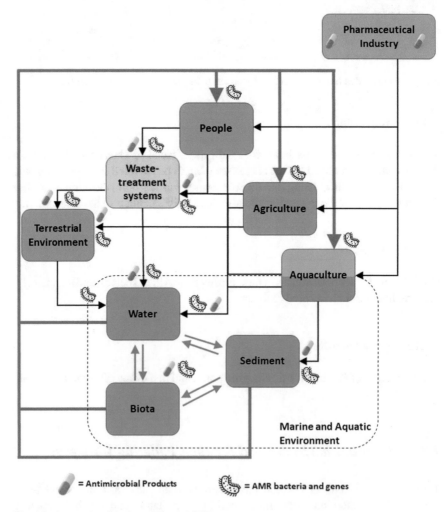

Fig. 1.2 Areas of antibiotics application [70]

when their metabolism consequent to resistant bacteria [71]. An existence of resistant bacteria causes a lot of problems [72–76] that are not only humanly hazardous, but environmentally dangerous as well.

1.3.3 Environmental Concerns and Health Issues Arise from PPs in Wastewater

There are mainly two negative impacts which become a matter of a concern. First, toxic elements from PPs become potential hazards to human lives, animals and plants. This becomes even more pronounced if mineralization of PPs is incomplete. Second, water pollution takes place from the biodegradation of PPs, and this pose an environmental concerns [77–80].

1.4 Wastewater

An adsorption of dyes and PPs into the wastewater cause a reflection of the sunlight into the wastewater, and that interferes with aquatic species growth and deters photosynthesis. These consequent involves both human and environmental hazards.

1.5 Photocatalyst Semiconductors

Photocatalysts semiconductors are fabricated predominantly form nanoparticles of metal oxides. The metal oxides mandatorily absorb the light under solar or direct light exposure. Under the application, photocatalysts semiconductors are termed either photoanode or photoelectrode. Photoanodes encompass suitable bandgaps which advantage to degradation of the wastewater pollutants [31].

1.6 Degradation Techniques

Wastewater degradation methods are divided into four categories: physical, chemical, biological, acoustical, radiation and electric processes [78]. Physical methods don't effectively remove pollutants, and causes secondary pollution. Biological methods also are problematic as organic pollutants become insufficiently removed from the wastewater. It also causes a release of carcinogenic compounds, and its process is long [78]. Hence, an application of photocatalysis technique became popular on wastewater treatment.

1.6.1 Photocatalysis Process

Photosynthesis technique is carried out either naturally or artificially. The natural process involves a utilization of the solar light, and the solar energy is created to produce electrical energy. Nonetheless, due to complications such as low energy, inconvenient storage and transportation, artificial photosynthesis process is necessitated. Under the artificial photosynthesis process, a direct sunlight is used to release hydrogen and oxygen imperative for the degradation of pollutants [79].

Synthetic dyes and pharmaceutical waste discussed thus far are the major pollutants of the wastewater. In fact, acidic and basic dyes have extensively been studied by numerous researchers in terms of the degradation efficiency on removal from the wastewater.

Saravanan et al. [80] worked on Sol-Gel hydrothermally synthesis and visible light photocatalytic degradation performance of Fe/N codoped TiO_2 catalysts on acidic orange 7 (AO7). From the analysis, maximum decolorization (more than 90%) was found when a combination of visible light irradiation and heterogeneous system was used on Fe/N–TiO_2 catalyst. The AO7 degradation was considerably improved by increasing the hydraulic retention time from 2.5 to 10 h or by reducing the initial AO7 concentration from 300 to 100 mg/l. The reaction rate increased with the light intensity and the maximum value occurred at 35 Mw/cm^2. Moreover, the efficiency of the AO7 degradation increased when the pH decreased with maximum the maximum efficiency at pH 3.

Karimi et al. [81] engaged a hydrothermally synthesized Bi–$Bi_2O_2CO_3$ heterojunction photo-anode to degrade MB using photocatalysis technique. Relatively, a pure $Bi_2O_2CO_3$ found to yield 60% degradation efficiency, while Bi–$Bi_2O_2CO_3$ heterojunction photo-anode significantly reached 100% degradation efficiency.

Rajamanickam et al. [82] compared photocatalytic performance of two synthesis photoactive semiconductors: mechano-thermally synthesized Fe/FeS nanostructures formed from micron-sized and thermally synthesized Fe/FeS produced from the nano-sized precursors. From the analysis, the following observations were acquired: Degradation of 5 ppm of the MB solution by mechano-thermally synthesized Fe/FeS with a photocatalyst density of 1 kg/m^3 at pH 11 reached 96% after 12 ks irradiation under visible light. Notably, the photocatalytic efficiency was noted to be higher in alkaline solution. In fact, the rate constant value from the thermally synthesized Fe/FeS photocatalyst sample was found to be 1.5 times better than that of the mechano-thermally synthesized material.

On utilization of Cu_2O nanobelts on nanoporous Cu substrate, the degradation of methylene blue (MB) and methylene orange (MO), Jo et al. [83] applied photocatalysis to determine the photocatalytic degradation efficiency between the two materials. Analysis from the two degradation efficiency revealed the higher degradation efficiency (95.4%) from the MO, and MB yielded to 93.6%.

As much as photocatalysis effectively remove pollutants from wastewater, it is known that it has two major drawbacks. These include a recombination rate of electron-hole pairs and inferior solar light absorption. All these lead to a low degradation efficiency on most applications. Hence, it is normally coordinated to electrochemical oxidation so that electrons from engaged semiconductor photocatalyst react with evolved oxygen to produce oxygen reactive radicals which react further with protons to generate H_2O^-. This creates radicals which undergo chemical reactions with the wastewater pollutants [49]. The combination of these methods is termed photo-electrochemical technique. The details on how the technique works can be found elsewhere [48].

Vivid significance of photoelectrochemical oxidation technique can be evidenced from the likes of Jo et al.'s work [83] where WO_3 nanoporous photocatalyst was utilized on photoelectrochemical oxidation of MO from the aqueous solution. An application of electrochemical oxidation and photocatalysis separately yielded to 33.4% and 70.4%, respectively. Significantly, the combined technique (photoelectrochemical) led to 95.7% of MO within 3 h of the photoelectrochemical degradation.

1.6.2   Photoelectrochemical Degradation Efficiency on Synthetic Dyes

Engaging of photoelectrochemical technique on wastewater treatment encompasses an immobilization of the composite fabricated on the photoelectrode through an application of a bias potential from the potentiostat. These results into a production of hydroxyl species, and then electrons from the photoelectrode react with released oxygen to develop oxygen reactive radicals which react with protons to generate H_2O^- imperatively ideal for a degradation of wastewater pollutants [48, 49].

From most recent studies on photoelectrochemical oxidation, the degradation efficiency of organic pollutants are observed to be hugely depending on a type of the semiconductor photocatalyst, synthesis of photocatalysts, parameters of synthetic method and the organic compound degraded.

On improvement of ZnO inherent electron emission propensity, Chakrabarti et al. [84] incorporated TiO_2 into the ZnO and characterize the impact of binary photocatalysts on degradation of MB. Within 180 min application of the Photocatalysts separately, TiO_2 gave degradation efficiency of 95.80%, while ZnO resulted to 38.68%. Significantly, the binary ZnO/TiO_2 reached to about 99.41% degradation efficiency of the MB.

Silva et al. [85] examined synthetic parameters' impact on photoelectrochemical degradation of the MO using TiO_2/Ti mesh photocathode. Their analysis revealed a better photoelectrochemical degradation efficiency (100%) from the TiO_2/Ti mesh photocathode optimum parameters of 550 °C, ph 1 and ratio composition of 20% rutile and 80% anatase.

Kyzas et al. [86] on their utilization of the Fe_2O_3 photocatalyst, produced by spray pyrolysis method, compared degradation efficiency between salicylic acid and

MO. Results revealed basic MO to be degraded to about 94% in just 160 min, while SA took 440 min to be degraded to 53%. Ahmed et al. [87] fabricated A/R TiO$_2$ composite for a photodegradation of orange II dye. A paramount photoelectrochemical parameter was found to be on anatase nanograins within TiO$_2$ nanorods, and 96% degradation efficiency was significantly obtained. On the other hand, Mirzaei et al. [88] developed TiO$_2$ nanotube arrays wrapped with g-C$_3$N$_4$ using both anodisation and impregnation-under-vacuum techniques to determine photoelectrochemical degradation of Phenol. Analysis revealed an inclusion of g-C$_3$N$_4$ to play a critical role on attained degradation efficiency over 90%.

Using hydrothermal method to produce Cu$_2$O nanowires as photocatalyst, Pathania et al. [89] discovered a 150 °C temperature as optimum for enhanced degradation efficiency of 90% within 20 min of photoelectrochemical degradation. Quite recently, Pathania et al. [89] utilized photoelectrochemical technique on TiO$_2$ nanorod produced by hydrothermal deposit. A high degradation efficiency was found from the TiO$_2$ nanorods deposited at TB-1.2 of titanium butoxide. Furthermore, 77% degradation efficiency was noted from Orange II, while 94% was found from the MB. These findings were attributed to aspect ratio, light trapping, excellent charge separation, etc.

From some of the included reported studies, it can be seen that the combination of photocatalysis and electrochemical oxidation lead to a better photocatalytic degradation under the variance of photocatalysists and synthetic dyes. Besides pin pointed degradation efficiency from the photoelectrochemical application, even the previously documented findings attest and factors are summarized in Table 1.1.

Table 1.1 Parameters leading to the dynamic degradation efficiency of organic compounds from wastewater

Photoanode	Organic compound	Synthetic technique	Degradation (%)	References
TiO$_2$	Malechite green	Atmospheric plasma	70–94	[90]
La/Nco-doped TiO$_2$	Malechite green	Sol-gel and deep coating	80	[91]
Pd/C–N–S–TiO$_2$	Acetylsalicylic acid (aspirin)	Galvanostatic method	90	[81]
Au–Fe$_2$O$_3$ thin films	SA	Spray pyrolysis technique	45	[92]
Fe$_2$O$_3$ thin films	MO	Liquid phase deposition	79.1	[70]
Fe$_2$O$_3$ thin films	MB	Hydrothermal synthesis	53.94	[73]

(continued)

Table 1.1 (continued)

Photoanode	Organic compound	Synthetic technique	Degradation (%)	References
Fe_2O_3 nanotubes	MB	Anodisation method	91	[72]
Fe_2O_3	SA	Spray pyrolysis method	53	[93]
Fe_2O_3	MO	Spray pyrolysis method	94	[94]
Fe_2O_3/Bi_2O_3 thin film	MO	Pore impregnation	64.5	[83]
Fe_2O_3 nanospheres	Rhodamine B	Simplified electrochemical route	90	[74]

1.7 Summary of the Findings

From the chapter outline, there are four pillars which are found to be critical on wastewater treatment, and they are summarized below.

1.7.1 Causative Impacts of Synthetic Dyes and Pharmaceutical Products Wastewater

On a present day vast industrial application of dyes and medications, synthetic dyes and pharmaceutical products are the main organic pollutants of the wastewater. Existence of these products in wastewater has a number of negative impacts. In fact, people, animal and plants exposed to excreted PPs could face a definitive health threads. Furthermore, excretion of PPs is also an environmental hazard.

1.7.2 Semiconductor Photocatalysts

Semiconductor photocatalysts naturally absorb light under solar application, they encompass suitable bandgaps to effectively photodegrade diversity of wastewater pollutants. Nonetheless, utilization of semiconductor photocatalysts under photocatalysis comprised inherent limitations. These are recombination rate of electron-hole pairs and inferior solar light absorption. Due to all these, photoelectrochemical became only feasible method.

1.7.3 Photoelectrochemical Technique

Photoelectrochemical technique consists an immobilization of the photocatalyst semiconductor on a conducting material with the application of a bias potential, and that lead to the generation of hydroxyl species. Furthermore, photo-electrons from photocatalyst semiconductor react with released oxygen to develop oxygen reactive radicals which react with protons to generate H_2O^-. Consequently, the radicals undergo chemical reactions with the wastewater pollutants for an effective degradation.

1.7.4 Photoelectrochemical Degradation Efficiency on Organic Pollutants

An application of photoelectrochemical technique has so far found to place significant inroads on wastewater degradation. This is particularly pronounced on degradation of various synthetic dyes. Importantly, four things are proved to be the main determining factors on photoelectrochemical degradation of synthetic dyes from the wastewater. These are: a type of the semiconductor photocatalyst, synthesis of photocatalysts, parameters of synthetic method and the organic compound degraded. From all these, dynamic degradation efficiency is obtained.

1.8 Conclusions

This chapter has described the most recent issues on removal of disposed pharmaceutical products and synthetic dyes from the wastewater. Dynamic degradation efficiency of various photocatalysts constituents has thoroughly been discussed and the following conclusions were drawn:

- Acid and basic dyes are more popular synthetic dyes, and that is due to their high-fastness, brilliance and versatility of application.
- Pharmaceutical products resulted into the wastewater from three major sources: hospitals, households and drug factories.
- Wastewater has toxic elements which have potential hazards to human lives, animals and plants. This becomes even more pronounced when mineralization of the organic pollutants is incomplete. Furthermore, wastewater encompasses biodegradation, and that negatively affects the environment.
- Semiconductor photocatalysts are predominantly synthesized from nanoparticles powders, and fabricated to photo-active semiconductors.

- Photocatalysis has inherent limitations of rapid photo-generation of electron-hole pairs, high operational costs and utilization of small portion of solar energy which renders to ineffective degradation.
- Synergy between photocatalysis and electrochemical oxidation is a true solution for the better degradation efficiency from utilization various photoelectrode constituents.
- Scruitinization on dynamic degradation efficiency, provided by photoelectrochemical technique on variety of organic pollutants proves that the degradation efficiency of organic pollutants hugely depends on: type of the semiconductor photocatalyst, synthesis of photocatalysts, parameters of synthetic method producing photocatalysts and the organic compound degraded.

References

1. P. Kar, T.K. Maji, R. Nandi, P. Lemmens, S.K. Pal, In-situ hydrothermal synthesis of Bi–$Bi_2O_2CO_3$ heterojunction photocatalyst with enhanced visible light photocatalytic activity. Nano-Micro Lett. **9**(2), 18 (2017)
2. H. Esmaili, A. Kotobi, S. Sheibani, F. Rashchi, Photocatalytic degradation of methylene blue by nanostructured Fe/FeS powder under visible light. Int. J. Miner. Metall. Mater. **25**(2), 244–252 (2018)
3. Y. Li, C. Ji, Y.C. Chi, Z.H. Dan, H.F. Zhang, F.X. Qin, Fabrication and photocatalytic activity of Cu_2O nanobelts on nanoporous Cu substrate. Acta Metallurgica Sinica (English Letters) **32**(1), 63–73 (2019)
4. H.H. Cheng, S.S. Chen, S.Y. Yang, H.M. Liu, K.S. Lin, Sol-gel hydrothermal synthesis and visible light photocatalytic degradation performance of Fe/N codoped TiO_2 catalysts. Materials **11**(6), 939 (2018)
5. F. Zhang, B. Zhang, X. Wang, L. Huang, D. Ji, S. Du et al., Synthesis of tetrahydropyran from tetrahydrofurfuryl alcohol over Cu–Zno/Al_2O_3 under a gaseous-phase condition. Catalysts **8**(3), 105 (2018)
6. C.S. Tseng, T. Wu, Y.W. Lin (2018). Facile synthesis and characterization of Ag_3PO_4 microparticles for degradation of organic dyestuffs under white-light light-emitting-diode irradiation. Materials **11**(5)
7. Y.J. Jang, C. Simer, T. Ohm, Comparison of zinc oxide nanoparticles and its nano-crystalline particles on the photocatalytic degradation of methylene blue. Mater. Res. Bull. **41**(1), 67–77 (2006)
8. T.J. Whang, M.T. Hsieh, H.H. Chen, Visible-light photocatalytic degradation of methylene blue with laser-induced Ag/ZnO nanoparticles. Appl. Surf. Sci. **258**(7), 2796–2801 (2012)
9. L.V. Trandafilović, D.J. Jovanović, X. Zhang, S. Ptasińska, M.D. Dramićanin, Enhanced photocatalytic degradation of methylene blue and methyl orange by ZnO: Eu nanoparticles. Appl. Catal. B **203**, 740–752 (2017)
10. O. Mekasuwandumrong, P. Pawinrat, P. Praserthdam, J. Panpranot, Effects of synthesis conditions and annealing post-treatment on the photocatalytic activities of ZnO nanoparticles in the degradation of methylene blue dye. Chem. Eng. J. **164**(1), 77–84 (2010)
11. G. Nagaraju, G.C. Shivaraju, G. Banuprakash, D. Rangappa, Photocatalytic activity of ZnO nanoparticles: synthesis via solution combustion method. Mater. Today: Proc. **4**(11), 11700–11705 (2017)
12. M.L. Fetterolf, H.V. Patel, J.M. Jennings, Adsorption of methylene blue and acid blue 40 on titania from aqueous solution. J. Chem. Eng. Data **48**(4), 831–835 (2003)

13. J. Tschirch, R. Dillert, D. Bahnemann, B. Proft, A. Biedermann, B. Goer, Photodegradation of methylene blue in water, a standard method to determine the activity of photocatalytic coatings? Res. Chem. Intermed. **34**(4), 381–392 (2008)

14. R.M. Epand, C. Walker, R.F. Epand, N.A. Magarvey, Molecular mechanisms of membrane targeting antibiotics. Biochimica et Biophysica Acta (BBA)-Biomembranes **1858**(5), 980–987 (2016)

15. S.N. Al-Bahry, I.Y. Mahmoud, M. Al-Zadjali, A. Elshafie, A. Al-Harthy, W. Al-Alawi, Antibiotic resistant bacteria as bio-indicator of polluted effluent in the green turtles, Chelonia mydas in Oman. Mar. Environ. Res. **71**(2), 139–144 (2011)

16. M. Vaara, New approaches in peptide antibiotics. Curr. Opin. Pharmacol. **9**(5), 571–576 (2009)

17. L.T. Nguyen, E.F. Haney, H.J. Vogel, The expanding scope of antimicrobial peptide structures and their modes of action. Trends Biotechnol. **29**(9), 464–472 (2011)

18. V. Vega-Sánchez, F. Latif-Eugenín, E. Soriano-Vargas, R. Beaz-Hidalgo, M.J. Figueras, M.G. Aguilera-Arreola, G. Castro-Escarpulli, Re-identification of Aeromonas isolates from rainbow trout and incidence of class 1 integron and β-lactamase genes. Vet. Microbiol. **172**(3–4), 528–533 (2014)

19. M. Mahboubi, G. Haghi, Antimicrobial activity and chemical composition of Mentha pulegium L. essential oil. J. Ethnopharmacol. **119**(2), 325–327 (2008)

20. R. Gothwal, T. Shashidhar, Antibiotic pollution in the environment: a review. Clean–Soil, Air, Water **43**(4), 479–489 (2015)

21. S.N. Al-Bahry, I.Y. Mahmoud, K.I.A. Al-Belushi, A.E. Elshafie, A. Al-Harthy, C.K. Bakheit, Coastal sewage discharge and its impact on fish with reference to antibiotic resistant enteric bacteria and enteric pathogens as bio-indicators of pollution. Chemosphere **77**(11), 1534–1539 (2009)

22. Y. Zhu, Y. Wang, Z. Chen, L. Qin, L. Yang, L. Zhu et al., Visible light induced photocatalysis on CdS quantum dots decorated TiO_2 nanotube arrays. Appl. Catal. A **498**, 159–166 (2015)

23. S.G. Kim, L.K. Dhandole, Y.S. Seo, H.S. Chung, W.S. Chae, M. Cho, J.S. Jang, Active composite photocatalyst synthesized from inactive Rh & Sb doped TiO_2 nanorods: enhanced degradation of organic pollutants & antibacterial activity under visible light irradiation. Appl. Catal. A **564**, 43–55 (2018)

24. M.E. Osugi, G.A. Umbuzeiro, M.A. Anderson, M.V.B. Zanoni, Degradation of metallophtalocyanine dye by combined processes of electrochemistry and photoelectrochemistry. Electrochim. Acta **50**(25–26), 5261–5269 (2005)

25. W. Liao, J. Yang, H. Zhou, M. Murugananthan, Y. Zhang, Electrochemically self-doped TiO_2 nanotube arrays for efficient visible light photoelectrocatalytic degradation of contaminants. Electrochim. Acta **136**, 310–317 (2014)

26. W. Li, F. Zhan, J. Li, C. Liu, Y. Yang, Y. Li, Q. Chen, Enhancing photoelectrochemical water splitting by aluminum-doped plate-like WO_3 electrodes. Electrochim. Acta **160**, 57–63 (2015)

27. X. Deng, Q. Ma, Y. Cui, X. Cheng, Q. Cheng, Fabrication of TiO_2 nanorods/nanosheets photoelectrode on Ti mesh by hydrothermal method for degradation of methylene blue: influence of calcination temperature. Appl. Surf. Sci. **419**, 409–417 (2017)

28. G.W. An, M.A. Mahadik, W.S. Chae, H.G. Kim, M. Cho, J.S. Jang, Enhanced solar photoelectrochemical conversion efficiency of the hydrothermally-deposited TiO_2 nanorod arrays: effects of the light trapping and optimum charge transfer. Appl. Surf. Sci. **440**, 688–699 (2018)

29. I.M. Szilágyi, B. Fórizs, O. Rosseler, Á. Szegedi, P. Németh, P. Király et al., WO_3 photocatalysts: influence of structure and composition. J. Catal. **294**, 119–127 (2012)

30. C.H. Li, C.W. Hsu, S.Y. Lu, TiO_2 nanocrystals decorated Z-schemed core-shell CdS-CdO nanorod arrays as high efficiency anodes for photoelectrochemical hydrogen generation. J. Colloid Interface Sci. **521**, 216–225 (2018)

31. H. Wu, Z. Zhang, Photoelectrochemical water splitting and simultaneous photoelectrocatalytic degradation of organic pollutant on highly smooth and ordered TiO_2 nanotube arrays. J. Solid State Chem. **184**(12), 3202–3207 (2011)

32. T.C. An, X.H. Zhu, Y. Xiong, Feasibility study of photoelectrochemical degradation of methylene blue with three-dimensional electrode-photocatalytic reactor. Chemosphere **46**(6), 897–903 (2002)

33. J. Zou, X. Peng, M. Li, Y. Xiong, B. Wang, F. Dong, B. Wang, Electrochemical oxidation of COD from real textile wastewaters: kinetic study and energy consumption. Chemosphere **171**, 332–338 (2017)

34. D. Liu, R. Tian, J. Wang, E. Nie, X. Piao, X. Li, Z. Sun, Photoelectrocatalytic degradation of methylene blue using F doped TiO_2 photoelectrode under visible light irradiation. Chemosphere **185**, 574–581 (2017)

35. R. Saravanan, V.K. Gupta, E. Mosquera, F. Gracia, V. Narayanan, A. Stephen, Visible light induced degradation of methyl orange using β-Ag0. $333V_2O_5$ nanorod catalysts by facile thermal decomposition method. J. Saudi Chem. Soc. **19**(5), 521–527 (2015)

36. P. Li, G. Zhao, K. Zhao, J. Gao, T. Wu, An efficient and energy saving approach to photocatalytic degradation of opaque high-chroma methylene blue wastewater by electrocatalytic pre-oxidation. Dyes Pigm. **92**(3), 923–928 (2012)

37. M.A. Mahadik, G.W. An, S. David, S.H. Choi, M. Cho, J.S. Jang, Fabrication of A/R-TiO_2 composite for enhanced photoelectrochemical performance: solar hydrogen generation and dye degradation. Appl. Surf. Sci. **426**, 833–843 (2017)

38. X. Yuan, J. Yi, H. Wang, H. Yu, S. Zhang, F. Peng, New route of fabricating BiOI and Bi_2O_3 supported TiO_2 nanotube arrays via the electrodeposition of bismuth nanoparticles for photocatalytic degradation of acid orange II. Mater. Chem. Phys. **196**, 237–244 (2017)

39. Y.M. Hunge, M.A. Mahadik, S.S. Kumbhar, V.S. Mohite, K.Y. Rajpure, N.G. Deshpande et al., Visible light catalysis of methyl orange using nanostructured WO_3 thin films. Ceram. Int. **42**(1), 789–798 (2016)

40. J. Tao, Z. Gong, G. Yao, Y. Cheng, M. Zhang, J. Lv et al., Enhanced photocatalytic and photoelectrochemical properties of TiO_2 nanorod arrays sensitized with CdS nanoplates. Ceram. Int. **42**(10), 11716–11723 (2016)

41. Y. Wang, J. Tao, X. Wang, Z. Wang, M. Zhang, G. He, Z. Sun, A unique Cu_2O/TiO_2 nanocomposite with enhanced photocatalytic performance under visible light irradiation. Ceram. Int. **43**(6), 4866–4872 (2017)

42. Z. Zainal, C.Y. Lee, M.Z. Hussein, A. Kassim, N.A. Yusof, Electrochemical-assisted photodegradation of mixed dye and textile effluents using TiO_2 thin films. J. Hazard. Mater. **146**(1–2), 73–80 (2007)

43. H. Wang, Y. Liang, L. Liu, J. Hu, W. Cui, Highly ordered TiO_2 nanotube arrays wrapped with g-C_3N_4 nanoparticles for efficient charge separation and increased photoelectrocatalytic degradation of phenol. J. Hazard. Mater. **344**, 369–380 (2018)

44. H.X. Han, C. Shi, L. Yuan, G.P. Sheng, Enhancement of methyl orange degradation and power generation in a photoelectrocatalytic microbial fuel cell. Appl. Energy **204**, 382–389 (2017)

45. M. Choi, J. Hwang, H. Setiadi, W. Chang, J. Kim, One-pot synthesis of molybdenum disulfide–reduced graphene oxide (MoS_2-RGO) composites and their high electrochemical performance as an anode in lithium ion batteries. J. Supercrit. Fluids **127**, 81–89 (2017)

46. G.G. Bessegato, J.C. Cardoso, M.V.B. Zanoni, Enhanced photoelectrocatalytic degradation of an acid dye with boron-doped TiO_2 nanotube anodes. Catal. Today **240**, 100–106 (2015)

47. H. Kmentova, S. Kment, L. Wang, S. Pausova, T. Vaclavu, R. Kuzel et al., Photoelectrochemical and structural properties of TiO_2 nanotubes and nanorods grown on FTO substrate: comparative study between electrochemical anodization and hydrothermal method used for the nanostructures fabrication. Catal. Today **287**, 130–136 (2017)

48. C. Fu, M. Li, H. Li, C. Li, X. Guo Wu, B. Yang, Fabrication of Au nanoparticle/TiO_2 hybrid films for photoelectrocatalytic degradation of methyl orange. J. Alloy. Compd. **692**, 727–733 (2017)

49. G.P. Awasthi, S.P. Adhikari, S. Ko, H.J. Kim, C.H. Park, C.S. Kim, Facile synthesis of ZnO flowers modified graphene like MoS2 sheets for enhanced visible-light-driven photocatalytic activity and antibacterial properties. J. Alloy. Compd. **682**, 208–215 (2016)

50. S.A. Ansari, M.M. Khan, M.O. Ansari, M.H. Cho, Silver nanoparticles and defect-induced visible light photocatalytic and photoelectrochemical performance of Ag@ m-TiO_2 nanocomposite. Sol. Energy Mater. Sol. Cells **141**, 162–170 (2015)

51. S.V. Mohite, V.V. Ganbavle, K.Y. Rajpure, Photoelectrochemical performance and photo-electrocatalytic degradation of organic compounds using Ga: WO_3 thin films. J. Photochem. Photobiol., A **344**, 56–63 (2017)
52. H. Ma, Q. Zhuo, B. Wang, Electro-catalytic degradation of methylene blue wastewater assisted by Fe_2O_3-modified kaolin. Chem. Eng. J. **155**(1–2), 248–253 (2009)
53. D. Liu, J. Zhou, J. Wang, R. Tian, X. Li, E. Nie et al., Enhanced visible light photoelectrocatalytic degradation of organic contaminants by F and Sn co-doped TiO_2 photoelectrode. Chem. Eng. J. **344**, 332–341 (2018)
54. Y.M. Hunge, A.A. Yadav, M.A. Mahadik, V.L. Mathe, C.H. Bhosale, A highly efficient visible-light responsive sprayed WO_3/FTO photoanode for photoelectrocatalytic degradation of brilliant blue. J. Taiwan Inst. Chem. Eng. **85**, 273–281 (2018)
55. C. Wu, Z. Gao, S. Gao, Q. Wang, H. Xu, Z. Wang et al., Ti3+ self-doped TiO_2 photoelectrodes for photoelectrochemical water splitting and photoelectrocatalytic pollutant degradation. J. Energy Chem. **25**(4), 726–733 (2016)
56. S.V. Mohite, V.V. Ganbavle, K.Y. Rajpure, Photoelectrocatalytic activity of immobilized Yb doped WO_3 photocatalyst for degradation of methyl orange dye. J. Energy Chem. **26**(3), 440–447 (2017)
57. M.M. Islam, S. Basu, Effect of morphology and pH on (photo) electrochemical degradation of methyl orange using TiO_2/Ti mesh photocathode under visible light. J. Environ. Chem. Eng. **3**(4), 2323–2330 (2015)
58. M.M. Islam, S. Basu, Understanding photoelectrochemical degradation of methyl orange using TiO_2/Ti mesh as photocathode under visible light. J. Environ. Chem. Eng. **4**(3), 3554–3561 (2016)
59. Y. Hunge, Photoelectrocatalytic degradation of methylene blue using spray deposited ZnO thin films under UV illumination. MOJ Poly. Sci. **1**(4), 00020 (2017)
60. P. Niu, D. Wang, A. Wang, Y. Liang, X. Wang, Fabrication of bifunctional TiO_2/POM microspheres using a layer-by-layer method and photocatalytic activity for methyl orange degradation. J. Nanomater. (2018)
61. N. Lezana, F. Fernández-Vidal, C. Berríos, E. Garrido-Ramírez, Electrochemical and photo-electrochemical processes of Methylene blue oxidation by Ti/TiO_2 electrodes modified with Fe-allophane. J. Chil. Chem. Soc. **62**(2), 3529–3534 (2017)
62. R.D. Suryavanshi, S.V. Mohite, A.A. Bagade, K.Y. Rajpure, Photoelectrocatalytic activity of immobilized Fe_2O_3 photoelectrode for degradation of salicylic acid and methyl orange dye under visible light illumination. Ionics **24**(6), 1841–1853 (2018)
63. Y. Ding, Y. Zhou, W. Nie, P. Chen, MoS2–GO nanocomposites synthesized via a hydrothermal hydrogel method for solar light photocatalytic degradation of methylene blue. Appl. Surf. Sci. **357**, 1606–1612 (2015)
64. Y.M. Hunge, V.S. Mohite, S.S. Kumbhar, K.Y. Rajpure, A.V. Moholkar, C.H. Bhosale, Photoelectrocatalytic degradation of methyl red using sprayed WO_3 thin films under visible light irradiation. J. Mater. Sci.: Mater. Electron. **26**(11), 8404–8412 (2015)
65. R.T. Sapkal, S.S. Shinde, T.R. Waghmode, S.P. Govindwar, K.Y. Rajpure, C.H. Bhosale, Photo-corrosion inhibition and photoactivity enhancement with tailored zinc oxide thin films. J. Photochem. Photobiol., B **110**, 15–21 (2012)
66. P. Chatchai, A.Y. Nosaka, Y. Nosaka, Photoelectrocatalytic performance of WO_3/BiVO$_4$ toward the dye degradation. Electrochim. Acta **94**, 314–319 (2013)
67. C. Yu, Y. Shu, X. Zhou, Y. Ren, Z. Liu, Multi-branched Cu_2O nanowires for photocatalytic degradation of methyl orange. Mater. Res. Exp. **5**(3), 035046 (2018)
68. Q. Zheng, C. Lee, Visible light photoelectrocatalytic degradation of methyl orange using anodized nanoporous WO_3. Electrochim. Acta **115**, 140–145 (2014)
69. N. Chaukura, W. Gwenzi, N. Tavengwa, M.M. Manyuchi, Biosorbents for the removal of synthetic organics and emerging pollutants: opportunities and challenges for developing countries. Environ. Dev. **19**, 84–89 (2016)
70. S. Garcia-Segura, S. Dosta, J.M. Guilemany, E. Brillas, Solar photoelectrocatalytic degradation of Acid Orange 7 azo dye using a highly stable TiO_2 photoanode synthesized by atmospheric plasma spray. Appl. Catal. B **132**, 142–150 (2013)

71. U.G. Akpan, B.H. Hameed, Parameters affecting the photocatalytic degradation of dyes using TiO$_2$-based photocatalysts: a review. J. Hazard. Mater. **170**(2–3), 520–529 (2009)
72. S.K. Tammina, B.K. Mandal, N.K. Kadiyala, Photocatalytic degradation of methylene blue dye by nonconventional synthesized SnO$_2$ nanoparticles. Environ. Nanotechnol. Monit. Manage. **10**, 339–350 (2018)
73. D. Cao, Y. Wang, X. Zhao, Combination of photocatalytic and electrochemical degradation of organic pollutants from water. Current Opin. Green Sustain. Chem. **6**, 78–84 (2017)
74. S. Garcia-Segura, E. Brillas, Applied photoelectrocatalysis on the degradation of organic pollutants in wastewaters. J. Photochem. Photobiol. C **31**, 1–35 (2017)
75. X. Meng, Z. Zhang, X. Li, Synergetic photoelectrocatalytic reactors for environmental remediation: a review. J. Photochem. Photobiol. C **24**, 83–101 (2015)
76. R.M. Fernández-Domene, R. Sánchez-Tovar, B. Lucas-granados, M.J. Muñoz-Portero, J. García-Antón, Elimination of pesticide atrazine by photoelectrocatalysis using a photoanode based on WO$_3$ nanosheets. Chem. Eng. J. (2018)
77. S. Ahmed, M.G. Rasul, R. Brown, M.A. Hashib, Influence of parameters on the heterogeneous photocatalytic degradation of pesticides and phenolic contaminants in wastewater: a short review. J. Environ. Manage. **92**(3), 311–330 (2011)
78. J. Saien, H. Nejati, Enhanced photocatalytic degradation of pollutants in petroleum refinery wastewater under mild conditions. J. Hazard. Mater. **148**(1–2), 491–495 (2007)
79. C. Hachem, F. Bocquillon, O. Zahraa, M. Bouchy, Decolourization of textile industry wastewater by the photocatalytic degradation process. Dyes Pigm. **49**(2), 117–125 (2001)
80. R. Saravanan, S. Karthikeyan, V.K. Gupta, G. Sekaran, V. Narayanan, A. Stephen, Enhanced photocatalytic activity of ZnO/CuO nanocomposite for the degradation of textile dye on visible light illumination. Mater. Sci. Eng., C **33**(1), 91–98 (2013)
81. L. Karimi, S. Zohoori, M.E. Yazdanshenas, Photocatalytic degradation of azo dyes in aqueous solutions under UV irradiation using nano-strontium titanate as the nanophotocatalyst. J. Saudi Chem. Soc. **18**(5), 581–588 (2014)
82. D. Rajamanickam, M. Shanthi, Photocatalytic degradation of an organic pollutant by zinc oxide–solar process. Arab. J. Chem. **9**, S1858–S1868 (2016)
83. W.K. Jo, R.J. Tayade, Recent developments in photocatalytic dye degradation upon irradiation with energy-efficient light emitting diodes. Chin. J. Catal. **35**(11), 1781–1792 (2014)
84. S. Chakrabarti, B.K. Dutta, Photocatalytic degradation of model textile dyes in wastewater using ZnO as semiconductor catalyst. J. Hazard. Mater. **112**(3), 269–278 (2004)
85. C.P. Silva, G. Jaria, M. Otero, V.I. Esteves, V. Calisto, Waste-based alternative adsorbents for the remediation of pharmaceutical contaminated waters: Has a step forward already been taken? Biores. Technol. **250**, 888–901 (2018)
86. G.Z. Kyzas, J. Fu, N.K. Lazaridis, D.N. Bikiaris, K.A. Matis, New approaches on the removal of pharmaceuticals from wastewaters with adsorbent materials. J. Mol. Liq. **209**, 87–93 (2015)
87. M.J. Ahmed, B.H. Hameed, Removal of emerging pharmaceutical contaminants by adsorption in a fixed-bed column: a review. Ecotoxicol. Environ. Saf. **149**, 257–266 (2018)
88. A. Mirzaei, Z. Chen, F. Haghighat, L. Yerushalmi, Removal of pharmaceuticals from water by homo/heterogonous Fenton-type processes–a review. Chemosphere **174**, 665–688 (2017)
89. D. Pathania, R. Katwal, G. Sharma, M. Naushad, M.R. Khan, H. Ala'a, Novel guar gum/Al$_2$O$_3$ nanocomposite as an effective photocatalyst for the degradation of malachite green dye. Int. J. Biol. Macromol. **87**, 366–374 (2016)
90. S. Arslan, M. Eyvaz, E. Gürbulak, E. Yüksel, A review of state-of-the-art technologies in dye-containing wastewater treatment–the textile industry case, in *Textile Wastewater Treatment* (InTech, 2016)
91. G. Elango, S.M. Roopan, Efficacy of SnO$_2$ nanoparticles toward photocatalytic degradation of methylene blue dye. J. Photochem. Photobiol. B **155**, 34–38 (2016)
92. N. Chaukura, E.C. Murimba, W. Gwenzi, Sorptive removal of methylene blue from simulated wastewater using biochars derived from pulp and paper sludge. Environ. Technol. Innovation **8**, 132–140 (2017)

93. N. Chaukura, B.B. Mamba, S.B. Mishra, Porous materials for the sorption of emerging organic pollutants from aqueous systems: the case for conjugated microporous polymers. J. Water Process Eng. **16**, 223–232 (2017)
94. V. Geissen, H. Mol, E. Klumpp, G. Umlauf, M. Nadal, M. van der Ploeg et al., Emerging pollutants in the environment: a challenge for water resource management. Int. Soil Water Conserv. Res. **3**(1), 57–65 (2015)

Chapter 2
Synthesis and Fabrication of Photoactive Nanocomposites Electrodes for the Degradation of Wastewater Pollutants

Onoyivwe Monday Ama, Khotso Khoele, David Jacobus Delport, Suprakas Sinha Ray, and Peter Ogbemudia Osifo

Abstract On the endeavor of wastewater treatment, synthesis and fabrication of various nanoparticle metal oxides is the pillar of continuation. As metal oxides possess different pros and cons, they are presently applied in their singular compounds and combined forms so as to pin point their definitive distinction on the efficiency and effectiveness occur on wastewater degradation. Hence, synthesis of metal oxides from nanoparticles and their fabrication to photoelectrode is paramount for clarification. Within this chapter, nanoparticles are explained; photoactive metal oxides are listed and discussed. Most centrally, the synthesis of photo-active semiconductors from nanoparticles and a fabrication of the photoactive metal oxides to semiconductor electrodes are detailed.

O. M. Ama (✉) · S. S. Ray
Department of Chemical Science, University of Johannesburg, Doornfontein, 2028, Johannesburg, South Africa
e-mail: onoyivwe4real@gmail.com

K. Khoele · D. J. Delport
Department of Chemical, Metallurgical and Materials Engineering, Tshwane University of Technology, Pretoria, South Africa

O. M. Ama · S. S. Ray
DST-CSIR National Center for Nanostructured Materials, Council for Scientific and Industrial Research, Pretoria 0001, South Africa

O. M. Ama · P. O. Osifo
Department of Chemical Engineering, Vaal University of Technology, Private Mail Bag X021, Vanderbijlpark 1900, South Africa

© Springer Nature Switzerland AG 2020   19
O. M. Ama and S. S. Ray (eds.), *Nanostructured Metal-Oxide Electrode Materials for Water Purification*, Engineering Materials,
https://doi.org/10.1007/978-3-030-43346-8_2

Graphical Abstract The purchased Nanoparticles are prepared as received through different steps. First, Nanoparticles are synthesized. Secondly, the intended Nanocomposites are formed using synthesis techniques such as sol-gel method. The formed Nanocomposite is characterized using different techniques. Thirdly, the Nanocomposite is pressed into a pellet. Fourthly, the copper wire is connected to the pellet using conductive silver paste. Finally, a transparent glass tube with opening at both ends sides was used.

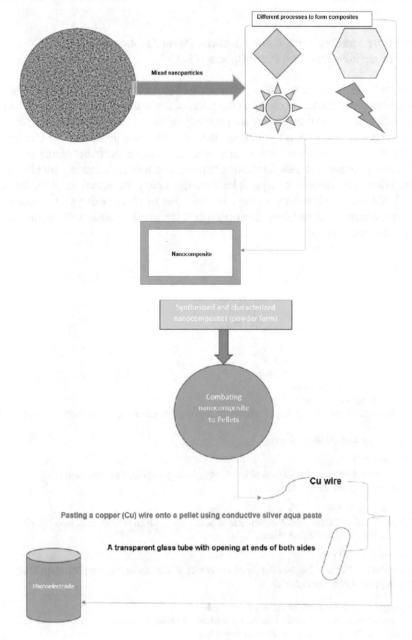

Highlights

- Photoactive metal oxides are listed and described.
- Nanoparticles from powders to nanocomposites are discussed.
- Synthesis of nanocomposites and fabrication of photoelectrodes are narrated upon.
- Current issues on wastewater treatment in terms of photocatalysists and degradation techniques are detailed.

2.1 Introduction

Diminishing of clean water and toxicity of wastewater increases rapidly from a plethora of an industrialization [1–10]. Industries such as textiles manufacturing paper printing, leathers making, ceramics making, cosmetics, inks and food processing have cause havoc on wastewater. A production of synthetic organic dyes is a stem and a center of all these industries. In fact, about 15% of organic dyes is lost annually into the wastewater [11–13]. Presence of dyes in wastewater leads to two distinct problems. First, toxic elements from dyes could lead to ill health of human beings, animals and plants. Second, dyes chemicals within wastewater negatively affect the environment [14–22]. All these necessitated abatement of organic dyes from the wastewater.

Due to a high surface to volume ratio and high catalytic activities [23–30], nanoparticles are used effectively in wastewater treatment. Nanoparticles for wastewater treatment should, however, be photoactive. Photoactive nanocomposites absorb light under solar application, and that is deal for the degradation of the organic pollutants from the wastewater. This significance is particularly enabled by their spontaneous appropriate bandgaps [31–38].

An application of photocatalyst semiconductors on wastewater treatment is carried by various methods, and photocatalysis technique once became the major one. However, photocatalysis have negative propensities on wastewater treatment. First, a recombination rate of electron-hole pairs occurs, and that affects effectiveness of engaged Photocatalyst. Second, it has inferior solar light absorption, and that consequent to lower degradation efficiency [39, 40]. Hence, the photocatalysis technique is incorporated to the electrochemical oxidation technique to form photoelectrochemical technique.

For the degradation of the wastewater, photoactive metal oxides from nanoparticles are synthesized to nanocomposites and then they are fabricated to photoelectrode semiconductors. In the interest of shedding some light on current issues on wastewater degradation, this chapter explains: nanoparticles, photoactive metal oxides, details on mandatory steps for the synthesis of nanocomposites from nanoparticles, and the fabrication of the photoelectrode semiconductors applied directly for the degradation of the wastewater pollutants.

2.2 Nanoparticles

2.2.1 Fundamental Nanoparticles

Nanoparticles are composed of powder particles which are less than hundred nanometers, and they are presently recognized to find diverse industrial application. Some of their predominance applications are on pharmaceutical, cosmetic, electronic, energy related projects and wastewater treatment [41]. This diversity of application comes from their two distinctive significances which include: high surface to volume ratio and high catalytic activities [42, 43].

Nanoparticles on wastewater treatment were initially applied through photocatalysis technique. However, due to a recombination rate of electron-hole pairs and inferior solar light absorption which affects effectiveness of Photoanodes when the photocatalysis technique was carried out [44], a coupling of electrochemical oxidation technique to photocatalysis was necessitated. The synergistic technique became photoelectrochemical technique.

2.2.2 Photoelectrochemical Technique

Photoelectrochemical technique composed an immobilization of the photoelectrode semiconductor onto a conducting substrate. This is carried out through an application of a bias potential. Then, electrons from the photoelectrodes react with released oxygen to develop oxygen reactive radicals which react with protons to generate H_2O^-. Finally; radicals undergo chemical reactions with the wastewater pollutants [45]. The whole process is accompanied by the generation of hydroxyl species from the combination of two techniques, and that significantly improves process efficiency and effectiveness of engaged photoanode [46]. In fact, the technique is now a favorable as it covers all inherent limitations of the photocatalysis on wastewater treatment. Figure 2.1 shows an overall set up of the photoelectrochemical technique, and the details therein can be found elsewhere [47].

2.3 Photoactive Nanocomposites

Nanocomposites are made up of metallic elements and rudiments such as oxygen, Sulphur, boron, etc. Fabricated nanocomposites are used in diverse space of applications such as on: wastewater treatment, coating for engineering components, surface-enhanced Raman scattering, Fluorescent and Surface Plasmon resonance. Distinctly, nanoparticles on application of wastewater treatment are only feasible with the use of photoactive nanoparticles.

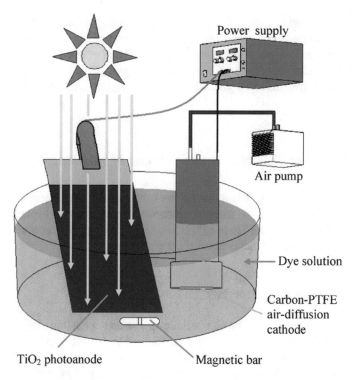

Fig. 2.1 Electrochemical cell composed of photoelectrochemical technique [47]

Photoactive nanocomposites spontaneously captivate light under the solar application, and their spontaneous bandgaps enable them to possess a high level of photodegradation efficiency on their application of removing pollutants from wastewater. In fact, photoactive nanocomposites have a stability and excellent charge carrier movement [48] ideal for the degradation. Hence, photoactive nanocomposites (photoanodes) are highly favorable for the photodegradation of wastewater pollutants [49–53].

Photoactive nanocomposites utilized for wastewater treatment are mostly in a form of metal oxides or chalcogenides such as MoS_2, ZnS and WS_2. Some of the mostly utilized metal oxides are TiO_2, MnO_2, WO_3, ZnO, Fe_2O_3, NiO, $SrZrO_3$, $BiOI/Ag_3VO_4$, $Ag_2O/Ag_3VO_4/AgVO_3$, etc. Listed photocatalysists above and the likes have been used on numerous studies on wastewater treatment [10, 23, 24, 54–65].

2.4 Synthesis of Photoactive Nanocomposites

Photoactive nanocomposites are produced from the nanoparticle reagents (powders) which are normally purchased from different companies [66–68]. The nanopowders are then synthesized to nanocomposites of interest through various techniques. Once the nanocomposite is obtained, it is taken to different characterization techniques which are engaged to ensure that powders are mixed and adhered well to each other. Hence, this section elaborates on different techniques which are used to produce nanocomposites from nanopowders. Their characterization on morphologies, phases, is included.

2.4.1 Different Techniques on Synthesis of Nanocomposites

Some of the mostly used techniques for the synthesis of nanocomposites include: hydrothermal process, sol-gel method, electrodeposition, chemical spray pyrolysis and Anodization technique. These methods are applied based on photocatalysis used and the organic pollutant that is under the degradation process.

2.4.1.1 Hydrothermal Process

Hydrothermal process (HP) is a dual technique with an application of aqueous solution and a higher temperature for the formation of the nanocomposite. The HP is normally carried within an autoclave with a presence of the catalyst. The typical HP process starts with nanoparticles suspended in the aqueous solution, and then an application of dynamic temperature. Thereafter, the product is filtered, washed and dried. Figure 2.2 shows the typical one step HP on the synthesis of ordered

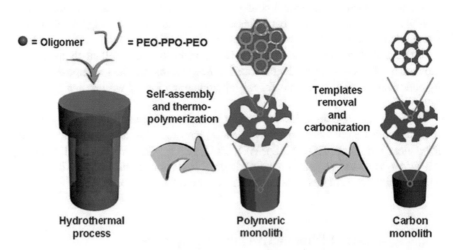

Fig. 2.2 One step HP of the ordered mesostructured carbonaceous monoliths, with hierarchical porosities [69]

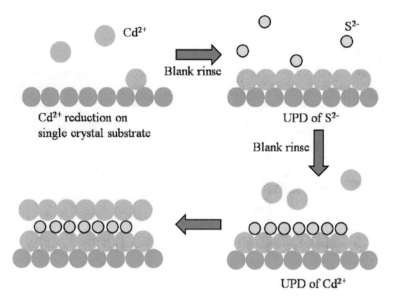

Fig. 2.3 Four steps engaged on the ED process [70]

mesostructured carbonaceous monoliths with hierarchical porosities. The details on the process can be found elsewhere [69].

2.4.1.2   Electrodeposition

Electrodeposition (ED) mechanical and electrical operation with photoactive nanoparticles on the classic ED application, a thin layer of the photoactive nanoparticles is deposited onto a contact layer of the substrate. A resultant film is then transferred to another substrate and the primary substrate is peeled off. Any of the excess powder is removed by ultrasonication in an aqueous solution. Finally, the synthesized nanocomposites can be used as the photoanode. Figure 2.3 shows all the steps engaged on the ED process [70]

2.4.1.3   Chemical Spray Pyrolysis Technique (CSP)

CSP involves the use of an aqueous solution for the deposition of nanocomposites particles from their aqueous phase to gaseous phase where they form a thin film layer. Figure 2.4 reveals a sketch of the whole CSP process. Nanoparticles on their liquid phase are chemically mixed with the water, and then the mixture is deposited on the substrate through a spraying nozzle. Thereafter, the substrate is heated up in an autoclave for some time. The CPS can be carried out either in one step or two steps [71].

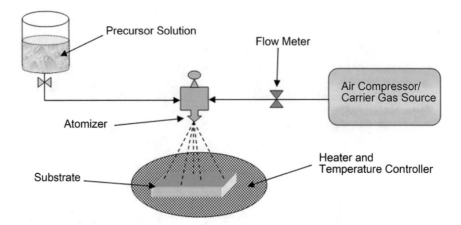

Fig. 2.4 Chemical spray pyrolysis technique applied on a substrate [71]

2.4.1.4 Anodization Technique

An Anodization method is carried out in an electrochemical cell comprised of the nanoparticles suspended in an aqueous solution, anode and cathode electrodes. Then, anodizing equipment operating at the certain voltage and current is applied to the cell. Henceforth, nanoparticles are electrochemically deposited onto anode electrode [72–75]. The anodized nanocomposite is then used in fabrication of photoactive nanocomposite electrode which is used for wastewater degradation. Figure 2.5 shows the Anodization process used on Ti wire for the formation of the Au and TiO_2 nanotubes.

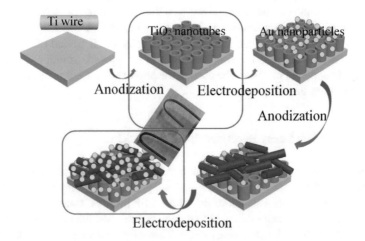

Fig. 2.5 Anodization technique for formation of TiO_2 and Au nanotubes [76]

2.4.1.5 Sol-Gel Method

Among discussed synthesis techniques, sol-gel has found popularity due to its numerous advantages and diversity. In fact, usage of the sol-gel has found to be going along with outstanding regulator of stoichiometry of most precursor solutions, affluence of compositional amendments, and flexibility on a microstructure formation, allows diversity of functional groups, with comparatively low annealing temperatures and allows coating deposition even on large-area substrates. Most of all, the whole process is relatively worthwhile. On the other hand, the sol-gel process is versatile, and it is enormously applied to form nanocomposites, fibers, monoliths, powder, spheres, grains, thin films (coating), etc. [77–80]. The versatility of the application is summarized in Fig. 2.6.

2.4.2 Characterization Techniques

Characterization techniques are engaged on powders, and they are also applied on nanocomposites before and after they have been synthesized and fabricated to photoelectrodes. Some of the characterization which is hugely utilized on nanocomposites are listed and explained below.

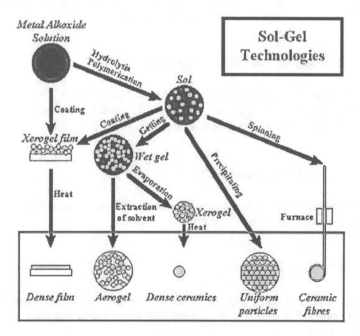

Fig. 2.6 Versatile applicable routes of the sol-gel process [81]

2.4.2.1 UV-Vis Absorbance Analysis

UV analysis uses wavelength as a function of the absorbance to examine homogeneity of the mixed nanocomposites. This is achieved by comparing a wavelength of each material with that of mixed degradation, the lesser the absorbance of the light the higher the degradation.

2.4.2.2 X-Ray Diffraction (XRD) and Raman Spectroscope

XRD is normally used to characterize crystalline formation of oxides on the surface of the substrate used, while Raman spectroscope is used to characterize organic material for the presence of D and G peaks. The D peak normally changes based on presence of certain species, while G peak confirms a response of stretching motion of symmetric bonds. In particular, the high intensity of the G band shows a presence of certain action. On the other hand, a low intensity of D band reveals very little disruption of the resultant action.

2.4.2.3 X-Ray Spectroscope (XPS)

This characterization technique is used to determine the chemical composition and elemental state of the engaged material after synthesis.

2.4.2.4 Scanning Electron Microscopy (SEM)

Mostly coupled to the energy dispersive X-Ray (EDX), SEM is primarily used to probe the surface morphologies of the samples in a solid or a powder form. The coupling of the EDX is for the elemental analysis.

2.4.2.5 Cyclic Voltammogram (CV)

CV is used mainly to compare current and peak-to-peak separation () from the drawn graph of current density versus potential.

2.4.2.6 BET Analysis

BET analysis are run from a nitrogen adsorption-desorption isotherm, and they are purposed to determine change which occurs on surface area and pore size of the engaged material.

2.4.2.7   Fourier-Transform Infrared Spectroscopy (FTIR)

Through a collection of a high spectral resolution data over a wide spectral range, FTIR is used to obtain an infrared spectrum of absorption or emission of a solid liquid samples.

2.5   Fabrication of the Photoactive Semiconductors

Synthesized nanocomposites are fabricated into the photoelectrodes which are used directly for the degradation of the wastewater. The process of photoelectrodes fabrication encompasses a compression of the nanocomposite to a pellet of a certain diameter. Then, a one end of a copper (Cu) wire is inserted and passed out through a glass tube with an opening at both ends and coiled onto the nanocomposite. Finally, a conduction between the Cu wire and the pellet is created with a use of conductive silver paste. For a proper close up on the glass tube, all the ends are covered up with an epoxy resin to prohibit water entrance.

2.6   Current Issues and Gap of Knowledge

Among variety of wastewater degradation techniques, photoelectrochemical technique is so far deemed to be superior [30, 70, 82–93]. Utilization of photoelectrochemical technique on wastewater treatment encompasses an immobilization of the composite fabricated on the photoelectrode through an application of a bias potential from the potentiostat. These results into a production of hydroxyl species, and then electrons from the photoelectrode react with released oxygen to develop oxygen reactive radicals which react with protons to generate H_2O^- imperatively ideal for a degradation of wastewater pollutants.

On account of continual studies for a complete removal (100%) of pollutants from the wastewater using photoelectrochemical technique, there are some factors which have been found to be problematic for an acquisition of the required photo activity on pertaining degradation process. Some of the most prevailing aspects include the following: semiconductor type, wide band gap energy and a high recombination of the electron-hole.

There are presently an extensive combination of n-type semiconductors and native p-type semiconductors together. The distinction between both types of semiconductors is that n-type material semiconductors have wider band gap energy, while p-types have narrowed band gap energy. Considering both types in favor of efficient photoelectrochemical degradation, and pros and cons each type has, they are combined together. For instance, ZnO as the n-type semiconductor has quite number of advantages such as: high surface reactivity, environment friendliness, cheap, etc. [70, 94–97] However, it is susceptible to photo-corrosion, and consequent to unstable photocatalytic activity. On the other hand, TiO_2 also as a good n-type semiconductor possess quite number of limitations [29, 98–106]. Chiefly, both ZnO and TiO_2 possess wider band gap energy.

The native p-type semiconductor like Cu_2O, Cds, CdSe, PbS, WO_3, etc. have narrowed band gap energy. On the actuality of Photocalysists which have different band gap energies, the semiconductors with narrow band gap energy are favorable for better degradation efficiency. Wide band gap negatively causes photocatalysts to have a high recombination of the electron-hole, and that consequent to a lower photocatalytic activity. Hence, both n-type and p-type are synergized for the better degradation efficiency. Figure 2.7 shows energy exchange when two semiconductors are mixed together.

On the other side of overcoming semiconductor with wide band gap energy, nanoparticles are used to coat semiconductors, and that enhances the photocatalytic activity of semiconductor and significantly degradation efficiency. Overall, present-day research route is on a type of the semiconductor used, method for synthesis of the nanocomposites, parameters of the synthetic method and the organic compound to be degraded.

The use of porous materials to immobilize nanocomposites on fabricated photoelectrodes is also predominating endeavor. One of the highly used porous materials is the exfoliated graphite (EG). This due to favorable properties which include: high thermal conductivity, good mechanical properties and excellent electrical conductivities, stable against aggressive media, high specific surface area, flexibility and compressibility, etc. Underpinning all these magnificent characteristics, quite number of works has been disseminated on significant impact of EG.

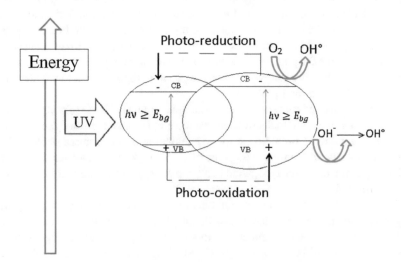

Fig. 2.7 Energy exchange when two different photocatalysists are combined together

2.7   Summary of the Chapter

Photocatalyst nanocomposites are critical for the degradation of organic pollutants from the wastewater. Within this chapter, synthesis of metal oxides from nanoparticles and their fabrication to photoelectrodes are detailed from photoanodes, methods of applications to the current issues of using photocatalyst nanocomposites.

Summary on the major part of this chapter include the following.

2.7.1   Nanoparticles

Nanoparticles are dimensionally structured to be less than hundred nanometers, and they are presently finding a diverse application on: engineering application for coatings, surface-enhanced Raman scattering, fluorescent and surface Plasmon resonance. They are also used for pharmaceutical, cosmetic, electronic, energy related and wastewater treatment.

2.7.2   Photoactive Metal Oxides

Photoactive metal oxides absorb light under solar application, and some of the mostly utilized ones are TiO_2, MnO_2, WO_3, ZnO, Fe_2O_3, NiO, $SrZrO_3$, $BiOI/Ag_3VO_4$, $Ag_2O/Ag_3VO_4/AgVO_3$, etc.

2.8   Methods for Production of Nanocomposites

Methods used on processing of nanocomposites are: hydrothermal process, sol-gel, electrodeposition, chemical spray pyrolysis, two-step chemical solution, one and two-pot hydrothermal and Anodization.

2.8.1   Characterization Techniques

Characterization techniques used on nanocomposites include: UV, SEM, XPS, XRD, FTIR, etc.

2.8.2   Synthesis of Photoactive Nanocomposites

Nanoparticle reagents (powders) are used on synthesis and fabrication of photoactive nanocomposites.

2.8.3   Fabrication of Photoelectrodes

Synthesized nanocomposites are fabricated into the photoelectrode semiconductors which are used directly engaged for the degradation of the wastewater. These processes of fabricating photoelectrodes encompass compression of the nanocomposites to a pellet of a certain diameter.

2.8.4   Photoelectrochemical Techniques

Photoelectrochemical technique is a voltage and light centered system which is applied onto the photoelectrode semiconductor through an immobilization of the photoelectrode.

2.9   Conclusions and the Future Direction

Synthesis of the metal oxides from nanoparticles and their fabrication to photoelectrode have, been discussed in this chapter. The most sensitized part of the chapter are on nanoparticles, photoactive metal oxides and photo-active semiconductors. From the critical center line of the study, the following conclusions have been drawn:

- Wastewaters have toxic elements which have potential hazards to human lives, animals, plants and the environment.
- Nanoparticle are versatile for industrial application, wastewater treatment is their vast area of the application.
- Nanoparticles as coating layer on semiconductors enhances their photocatalytic activity, and significantly degradation efficiency.
- Semiconductor photocatalysts are predominantly synthesized from nanoparticles powders, and are fabricated to photo-active semiconducting photoelectrodes.
- Photoelectrochemical technique yields better degradation efficiency from the wastewater treatment.
- Semiconductors with narrowed band gap energy are favorable for better degradation efficiency, and they are mixed with n-type semiconductors for reduction of their wide bandgaps.

- The use of porous materials plays a vital role on stability of Photocalysists and improves efficiency of photoelectrochemical technique.

References

1. P. Li, G. Zhao, K. Zhao, J. Gao, T. Wu, An efficient and energy saving approach to photo-catalytic degradation of opaque high-chroma methylene blue wastewater by electrocatalytic pre-oxidation. Dyes Pigm. **92**(3), 923–928 (2012)
2. W.K. Jo, R.J. Tayade, Recent developments in photocatalytic dye degradation upon irradiation with energy-efficient light emitting diodes. Chin. J. Catal. **35**(11), 1781–1792 (2014)
3. N. Chaukura, W. Gwenzi, N. Tavengwa, M.M. Manyuchi, Biosorbents for the removal of synthetic organics and emerging pollutants: opportunities and challenges for developing countries. Environ. Dev. **19**, 84–89 (2016)
4. N. Chaukura, B.B. Mamba, S.B. Mishra, Porous materials for the sorption of emerging organic pollutants from aqueous systems: the case for conjugated microporous polymers. J. Water Process Eng. **16**, 223–232 (2017)
5. V. Geissen, H. Mol, E. Klumpp, G. Umlauf, M. Nadal, M. van der Ploeg et al., Emerging pollutants in the environment: a challenge for water resource management. Int. Soil Water Conserv. Res. **3**(1), 57–65 (2015)
6. J.T. Jasper, O.S. Shafaat, M.R. Hoffmann, Electrochemical transformation of trace organic contaminants in latrine wastewater. Environ. Sci. Technol. **50**(18), 10198–10208 (2016)
7. F.C. Moreira, R.A. Boaventura, E. Brillas, V.J. Vilar, Electrochemical advanced oxidation processes: a review on their application to synthetic and real wastewaters. Appl. Catal. B **202**, 217–261 (2017)
8. S. Garcia-Segura, E. Brillas, Applied photoelectrocatalysis on the degradation of organic pollutants in wastewaters. J. Photochem. Photobiol., C **31**, 1–35 (2017)
9. S. Natarajan, H.C. Bajaj, R.J. Tayade, Recent advances based on the synergetic effect of adsorption for removal of dyes from waste water using photocatalytic process. J. Environ. Sci. **65**, 201–222 (2018)
10. R.M. Fernández-Domene, R. Sánchez-Tovar, B. Lucas-granados, M.J. Munoz-Portero, J. García-Antón, Elimination of pesticide atrazine by photoelectrocatalysis using a photoanode based on WO_3 nanosheets. Chem. Eng. J. **350**, 1114–1124 (2018)
11. A.M.S. Solano, C.A. Martínez-Huitle, S. Garcia-Segura, A. El-Ghenymy, E. Brillas, Application of electrochemical advanced oxidation processes with a boron-doped diamond anode to degrade acidic solutions of Reactive Blue 15 (Turqueoise Blue) dye. Electrochim. Acta **197**, 210–220 (2016)
12. X.L. He, C. Song, Y.Y. Li, N. Wang, L. Xu, X. Han, D.S. Wei, Efficient degradation of azo dyes by a newly isolated fungus Trichoderma tomentosum under non-sterile conditions. Ecotoxicol. Environ. Saf. **150**, 232–239 (2018)
13. P. Nigam, I.M. Banat, D. Singh, R. Marchant, Microbial process for the decolorization of textile effluent containing azo, diazo and reactive dyes. Process Biochem. **31**(5), 435–442 (1996)
14. S.K. Tammina, B.K. Mandal, N.K. Kadiyala, Photocatalytic degradation of methylene blue dye by nonconventional synthesized SnO_2 nanoparticles. Environ. Nanotechnol. Monit. Manage. **10**, 339–350 (2018)
15. A. Mirzaei, Z. Chen, F. Haghighat, L. Yerushalmi, Removal of pharmaceuticals from water by homo/heterogonous Fenton-type processes—a review. Chemosphere **174**, 665–688 (2017)
16. M.J. Ahmed, B.H. Hameed, Removal of emerging pharmaceutical contaminants by adsorption in a fixed-bed column: a review. Ecotoxicol. Environ. Saf. **149**, 257–266 (2018)

17. S. Arslan, M. Eyvaz, E. Gürbulak, E. Yüksel, A review of state-of-the-art technologies in dye-containing wastewater treatment–the textile industry case, in *Textile Wastewater Treatment*. InTech (2016)

18. A. El-Ghenymy, F. Centellas, J.A. Garrido, R.M. Rodríguez, I. Sirés, P.L. Cabot, E. Brillas, Decolorization and mineralization of Orange G azo dye solutions by anodic oxidation with a boron-doped diamond anode in divided and undivided tank reactors. Electrochim. Acta **130**, 568–576 (2014)

19. X. Florenza, S. Garcia-Segura, F. Centellas, E. Brillas, Comparative electrochemical degradation of salicylic and aminosalicylic acids: influence of functional groups on decay kinetics and mineralization. Chemosphere **154**, 171–178 (2016)

20. H.H. Ngo, W. Guo, J. Zhang, S. Liang, C. Ton-That, X. Zhang, Typical low cost biosorbents for adsorptive removal of specific organic pollutants from water. Biores. Technol. **182**, 353–363 (2015)

21. E. do Vale-Júnior, S. Dosta, I.G. Cano, J.M. Guilemany, S. Garcia-Segura, C.A. Martínez-Huitle, Acid blue 29 decolorization and mineralization by anodic oxidation with a cold gas spray synthesized Sn–Cu–Sb alloy anode. Chemosphere **148**, 47–54 (2016)

22. C.P. Silva, G. Jaria, M. Otero, V.I. Esteves, V. Calisto, Waste-based alternative adsorbents for the remediation of pharmaceutical contaminated waters: has a step forward already been taken? Biores. Technol. **250**, 888–901 (2018)

23. X. Meng, Z. Zhang, X. Li, Synergetic photoelectrocatalytic reactors for environmental remediation: a review. J. Photochem. Photobiol., C **24**, 83–101 (2015)

24. S. Garcia-Segura, S. Dosta, J.M. Guilemany, E. Brillas, Solar photoelectrocatalytic degradation of Acid Orange 7 azo dye using a highly stable TiO_2 photoanode synthesized by atmospheric plasma spray. Appl. Catal. B **132**, 142–150 (2013)

25. S.K. Brar, M. Verma, R.D. Tyagi, R.Y. Surampalli, Engineered nanoparticles in wastewater and wastewater sludge—evidence and impacts. Waste Manag. **30**(3), 504–520 (2010)

26. X. Qu, P.J. Alvarez, Q. Li, Applications of nanotechnology in water and wastewater treatment. Water Res. **47**(12), 3931–3946 (2013)

27. S. Olivera, H.B. Muralidhara, K. Venkatesh, V.K. Guna, K. Gopalakrishna, Y. Kumar, Potential applications of cellulose and chitosan nanoparticles/composites in wastewater treatment: a review. Carbohyd. Polym. **153**, 600–618 (2016)

28. J. Brame, Q. Li, P.J. Alvarez, Nanotechnology-enabled water treatment and reuse: emerging opportunities and challenges for developing countries. Trends Food Sci. Technol. **22**(11), 618–624 (2011)

29. E.H. Umukoro, M.G. Peleyeju, J.C. Ngila, O.A. Arotiba, Towards wastewater treatment: photo-assisted electrochemical degradation of 2-nitrophenol and orange II dye at a tungsten trioxide-exfoliated graphite composite electrode. Chem. Eng. J. **317**, 290–301 (2017)

30. M.M. Islam, S. Basu, Understanding photoelectrochemical degradation of methyl orange using TiO_2/Ti mesh as photocathode under visible light. J. Environ. Chem. Eng. **4**(3), 3554–3561 (2016)

31. S.K. Tammina, B.K. Mandal, N.K. Kadiyala, Photocatalytic degradation of methylene blue dye by nonconventional synthesized SnO_2 nanoparticles. Environ. Nanotechnol. Monit. Manag. **10**, 339–350 (2018)

32. S. Malato, P. Fernández-Ibáñez, M.I. Maldonado, J. Blanco, W. Gernjak, Decontamination and disinfection of water by solar photocatalysis: recent overview and trends. Catal. Today **147**(1), 1–59 (2009)

33. I. Poulios, D. Makri, X. Prohaska, Photocatalytic treatment of olive milling waste water: oxidation of protocatechuic acid. Global Nest: Int. J. **1**(1), 55–62 (1999)

34. B.K. Koo, D.Y. Lee, H.J. Kim, W.J. Lee, J.S. Song, H.J. Kim, Seasoning effect of dye-sensitized solar cells with different counter electrodes. J. Electroceram. **17**(1), 79–82 (2006)

35. J. Luo, M. Hepel, Photoelectrochemical degradation of naphthol blue black diazo dye on WO_3 film electrode. Electrochim. Acta **46**(19), 2913–2922 (2001)

36. D. Pathania, R. Katwal, G. Sharma, M. Naushad, M.R. Khan, H. Ala'a, Novel guar gum/Al_2O_3 nanocomposite as an effective photocatalyst for the degradation of malachite green dye. Int. J. Biol. Macromol. **87**, 366–374 (2016)

37. V.M. Daskalaki, M. Antoniadou, G. Li Puma, D.I. Kondarides, P. Lianos, Solar light-responsive Pt/CdS/TiO$_2$ photocatalysts for hydrogen production and simultaneous degradation of inorganic or organic sacrificial agents in wastewater. Environ. Sci. Technol. **44**(19), 7200–7205 (2010)

38. G. Elango, S.M. Roopan, Efficacy of SnO$_2$ nanoparticles toward photocatalytic degradation of methylene blue dye. J. Photochem. Photobiol., B **155**, 34–38 (2016)

39. Y. Ding, Y. Zhou, W. Nie, P. Chen, MoS$_2$–GO nanocomposites synthesized via a hydrothermal hydrogel method for solar light photocatalytic degradation of methylene blue. Appl. Surf. Sci. **357**, 1606–1612 (2015)

40. I.M. Szilágyi, B. Fórizs, O. Rosseler, Á. Szegedi, P. Németh, P. Király et al., WO$_3$ photocatalysts: influence of structure and composition. J. Catal. **294**, 119–127 (2012)

41. H. Ma, Q. Zhuo, B. Wang, Electro-catalytic degradation of methylene blue wastewater assisted by Fe$_2$O$_3$-modified kaolin. Chem. Eng. J. **155**(1–2), 248–253 (2009)

42. B. Jiang, L. Jiang, X. Shi, W. Wang, G. Li, F. Zhu, D. Zhang, Ag$_2$O/TiO$_2$ nanorods heterojunctions as a strong visible-light photocatalyst for phenol treatment. J. Sol-Gel. Sci. Technol. **73**(2), 314–321 (2015)

43. X. Li, H. Xu, W. Yan, Fabrication and characterization of PbO$_2$ electrode modified with polyvinylidene fluoride (PVDF). Appl. Surf. Sci. **389**, 278–286 (2016)

44. L. Wang, H. Zhai, G. Jin, X. Li, C. Dong, H. Zhang et al., 3D porous ZnO–SnS p–n heterojunction for visible light driven photocatalysis. Phys. Chem. Chem. Phys. **19**(25), 16576–16585 (2017)

45. J. Zhang, Z. Xiong, X.S. Zhao, Graphene–metal–oxide composites for the degradation of dyes under visible light irradiation. J. Mater. Chem. **21**(11), 3634–3640 (2011)

46. E. Vasilaki, M. Vamvakaki, N. Katsarakis, Complex ZnO-TiO$_2$ core-shell flower-like architectures with enhanced photocatalytic performance and superhydrophilicity without UV irradiation. Langmuir **34**(31), 9122–9132 (2018)

47. Q. Li, F. Wang, L. Sun, Z. Jiang, T. Ye, M. Chen et al., Design and synthesis of Cu@ CuS yolk–shell structures with enhanced photocatalytic activity. Nano-micro Lett. **9**(3), 35 (2017)

48. H.H. Cheng, S.S. Chen, S.Y. Yang, H.M. Liu, K.S. Lin, Sol-gel hydrothermal synthesis and visible light photocatalytic degradation performance of Fe/N codoped TiO$_2$ catalysts. Materials **11**(6), 939 (2018)

49. Y. Li, C. Ji, Y.C. Chi, Z.H. Dan, H.F. Zhang, F.X. Qin, Fabrication and photocatalytic activity of Cu$_2$O nanobelts on nanoporous Cu substrate. Acta Metall. Sin. (Eng. Lett.) **32**(1), 63–73 (2019)

50. M.A.B. Adnan, K. Arifin, L.J. Minggu, M.B. Kassim, Titanate-based perovskites for photochemical and photoelectrochemical water splitting applications: a review. Int. J. Hydr. Energy (2018)

51. C. Wu, Z. Gao, S. Gao, Q. Wang, H. Xu, Z. Wang et al., Ti$_3$+ self-doped TiO$_2$ photoelectrodes for photoelectrochemical water splitting and photoelectrocatalytic pollutant degradation. J. Energy Chem. **25**(4), 726–733 (2016)

52. S.A. Ansari, M.M. Khan, M.O. Ansari, M.H. Cho, Silver nanoparticles and defect-induced visible light photocatalytic and photoelectrochemical performance of Ag@ m-TiO$_2$ nanocomposite. Sol. Energy Mater. Sol. Cells **141**, 162–170 (2015)

53. G.W. An, M.A. Mahadik, W.S. Chae, H.G. Kim, M. Cho, J.S. Jang, Enhanced solar photoelectrochemical conversion efficiency of the hydrothermally-deposited TiO$_2$ nanorod arrays: effects of the light trapping and optimum charge transfer. Appl. Surf. Sci. **440**, 688–699 (2018)

54. Y. Huang, Y. Cai, H. Feng, D. Gu, Y. Deng, B. Tu et al., One-step hydrothermal synthesis of ordered mesostructured carbonaceous monoliths with hierarchical porosities. Chem. Commun. **23**, 2641–2643 (2008)

55. A. Ray. Electrodeposition of thin films for low-cost solar cells, in *Electroplating of Nanostructures*. IntechOpen (2015)

56. O.J. Ilegbusi, S.N. Khatami, L.I. Trakhtenberg, Spray pyrolysis deposition of single and mixed oxide thin films. Mater. Sci. Appl. **8**(02), 153 (2017)

57. N. Liu, S.P. Albu, K. Lee, S. So, P. Schmuki, Water annealing and other low temperature treatments of anodic TiO_2 nanotubes: a comparison of properties and efficiencies in dye sensitized solar cells and for water splitting. Electrochim. Acta **82**, 98–102 (2012)
58. K. Nakata, A. Fujishima, TiO_2 photocatalysis: design and applications. J. Photochem. Photobiol., C **13**(3), 169–189 (2012)
59. S. Li, J. Qiu, M. Ling, F. Peng, B. Wood, S. Zhang, Photoelectrochemical characterization of hydrogenated TiO_2 nanotubes as photoanodes for sensing applications. ACS Appl. Mater. Interfaces. **5**(21), 11129–11135 (2013)
60. J.H. Pan, H. Dou, Z. Xiong, C. Xu, J. Ma, X.S. Zhao, Porous photocatalysts for advanced water purifications. J. Mater. Chem. **20**(22), 4512–4528 (2010)
61. C. Fu, M. Li, H. Li, C. Li, X. Guo Wu, B. Yang, Fabrication of Au nanoparticle/TiO_2 hybrid films for photoelectrocatalytic degradation of methyl orange. J. Alloy. Compd. **692**, 727–733 (2017)
62. D. Liu, J. Zhou, J. Wang, R. Tian, X. Li, E. Nie et al., Enhanced visible light photoelectro-catalytic degradation of organic contaminants by F and Sn co-doped TiO_2 photoelectrode. Chem. Eng. J. **344**, 332–341 (2018)
63. D. Liu, R. Tian, J. Wang, E. Nie, X. Piao, X. Li, Z. Sun, Photoelectrocatalytic degrada-tion of methylene blue using F doped TiO_2 photoelectrode under visible light irradiation. Chemosphere **185**, 574–581 (2017)
64. D. Cao, Y. Wang, X. Zhao, Combination of photocatalytic and electrochemical degradation of organic pollutants from water. Current Opin. Green Sustain. Chem. **6**, 78–84 (2017)
65. J. Tao, Z. Gong, G. Yao, Y. Cheng, M. Zhang, J. Lv et al., Enhanced photocatalytic and photoelectrochemical properties of TiO_2 nanorod arrays sensitized with CdS nanoplates. Ceram. Int. **42**(10), 11716–11723 (2016)
66. E.V. dos Santos, O. Scialdone, Photo-electrochemical technologies for removing organic compounds in wastewater, in *Electrochemical Water and Wastewater Treatment* (2018), pp. 239–266
67. S.H.S. Chan, T. Yeong Wu, J.C. Juan, C.Y. Teh, Recent developments of metal oxide semi-conductors as photocatalysts in advanced oxidation processes (AOPs) for treatment of dye waste-water. J. Chem. Technol. Biotechnol. **86**(9), 1130–1158 (2011)
68. R.M. Asmussen, M. Tian, A. Chen, A new approach to wastewater remediation based on bifunctional electrodes. Environ. Sci. Technol. **43**(13), 5100–5105 (2009)
69. G.R.P. Malpass, D.W. Miwa, A.C.P. Miwa, S.A.S. Machado, A.J. Motheo,. Photo-assisted electrochemical oxidation of atrazine on a commercial Ti/Ru0. 3Ti0. 7O2 DSA electrode. Environ. Sci. Technol. **41**(20), 7120–7125 (2007)
70. O.M. Ama, N. Mabuba, O.A. Arotiba, Synthesis, characterization, and application of exfo-liated graphite/zirconium nanocomposite electrode for the photoelectrochemical degradation of organic dye in water. Electrocatalysis **6**(4), 390–397 (2015)
71. Y.M. Hunge, Photoelectrocatalytic degradation of methylene blue using spray deposited ZnO thin films under UV illumination. MO J. Polym. Sci. **1**, 00020 (2017)
72. Y.M. Hunge, A.A. Yadav, M.A. Mahadik, V.L. Mathe, C.H. Bhosale, A highly efficient visible-light responsive sprayed WO_3/FTO photoanode for photoelectrocatalytic degradation of brilliant blue. J. Taiwan Inst. Chem. Eng. **85**, 273–281 (2018)
73. N. Lezana, F. Fernández-Vidal, C. Berríos, E. Garrido-Ramírez, Electrochemical and photo-electrochemical processes of methylene blue oxidation by Ti/TiO_2 electrodes modified with Fe-allophane. J. Chil. Chem. Soc. **62**(2), 3529–3534 (2017)
74. Y.J. Jang, C. Simer, T. Ohm, Comparison of zinc oxide nanoparticles and its nano-crystalline particles on the photocatalytic degradation of methylene blue. Mater. Res. Bull. **41**(1), 67–77 (2006)
75. O. Mekasuwandumrong, P. Pawinrat, P. Praserthdam, J. Panpranot, Effects of synthesis con-ditions and annealing post-treatment on the photocatalytic activities of ZnO nanoparticles in the degradation of methylene blue dye. Chem. Eng. J. **164**(1), 77–84 (2010)
76. G. Nagaraju, G.C. Shivaraju, G. Banuprakash, D. Rangappa, Photocatalytic activity of ZnO nanoparticles: synthesis via solution combustion method. Mater. Today: Proc. **4**(11), 11700–11705 (2017)

77. L.V. Trandafilović, D.J. Jovanović, X. Zhang, S. Ptasińska, M.D. Dramićanin, Enhanced photocatalytic degradation of methylene blue and methyl orange by ZnO: Eu nanoparticles. Appl. Catal. B **203**, 740–752 (2017)
78. M. Brzezińska, P. García-Muñoz, A. Ruppert, N. Keller, Photoactive ZnO materials for solar light-induced CuxO–ZnO catalyst preparation. Materials **11**(11), 2260 (2018)
79. Z.B. Wang, H.X. Hu, Y.G. Zheng, W. Ke, Y.X. Qiao, Comparison of the corrosion behavior of pure titanium and its alloys in fluoride-containing sulfuric acid. Corros. Sci. **103**, 50–65 (2016)
80. U.G. Akpan, B.H. Hameed, Parameters affecting the photocatalytic degradation of dyes using TiO_2-based photocatalysts: a review. J. Hazard. Mater. **170**(2–3), 520–529 (2009)
81. X. Ma, Z. Sun, X. Hu, Synthesis of tin and molybdenum co-doped TiO_2 nanotube arrays for the photoelectrocatalytic oxidation of phenol in aqueous solution. Mater. Sci. Semicond. Process. **85**, 150–159 (2018)
82. J. Tschirch, R. Dillert, D. Bahnemann, B. Proft, A. Biedermann, B. Goer, Photodegradation of methylene blue in water, a standard method to determine the activity of photocatalytic coatings? Res. Chem. Intermed. **34**(4), 381–392 (2008)
83. W. Liao, J. Yang, H. Zhou, M. Murugananthan, Y. Zhang, Electrochemically self-doped TiO_2 nanotube arrays for efficient visible light photoelectrocatalytic degradation of contaminants. Electrochim. Acta **136**, 310–317 (2014)
84. C. Wang, F. Wang, M. Xu, C. Zhu, W. Fang, Y. Wei, Electrocatalytic degradation of methylene blue on Co doped Ti/TiO_2 nanotube/PbO_2 anodes prepared by pulse electrodeposition. J. Electroanal. Chem. **759**, 158–166 (2015)
85. P. Prasannalakshmi, N. Shanmugam, Phase-dependant photochemistry of TiO_2 nanoparticles in the degradation of organic dye methylene blue under solar light irradiation. Appl. Phys. A **123**(9), 586 (2017)
86. S.G. Kim, L.K. Dhandole, Y.S. Seo, H.S. Chung, W.S. Chae, M. Cho, J.S. Jang, Active composite photocatalyst synthesized from inactive Rh & Sb doped TiO_2 nanorods: Enhanced degradation of organic pollutants & antibacterial activity under visible light irradiation. Appl. Catal. A **564**, 43–55 (2018)
87. S.V. Mohite, V.V. Ganbavle, K.Y. Rajpure, Photoelectrocatalytic activity of immobilized Yb doped WO_3 photocatalyst for degradation of methyl orange dye. J. Energy Chem. **26**(3), 440–447 (2017)
88. P. Niu, D. Wang, A. Wang, Y. Liang, X. Wang, Fabrication of bifunctional TiO_2/POM microspheres using a layer-by-layer method and photocatalytic activity for methyl orange degradation. J. Nanomater. (2018)
89. R. Daghrir, P. Drogui, D. Robert, Photoelectrocatalytic technologies for environmental applications. J. Photochem. Photobiol., A **238**, 41–52 (2012)
90. C.S. Tseng, T. Wu, Y.W. Lin, Facile synthesis and characterization of Ag_3PO_4 microparticles for degradation of organic dyestuffs under white-light light-emitting-diode irradiation. Materials **11**(5), 708 (2018)
91. T.J. Whang, M.T. Hsieh, H.H. Chen, Visible-light photocatalytic degradation of methylene blue with laser-induced Ag/ZnO nanoparticles. Appl. Surf. Sci. **258**(7), 2796–2801 (2012)
92. O.M. Ama, O.A. Arotiba, Exfoliated graphite/titanium dioxide for enhanced photoelectrochemical degradation of methylene blue dye under simulated visible light irradiation. J. Electroanal. Chem. **803**, 157–164 (2017)
93. O.M. Ama, N. Kumar, F.V. Adams, S.S. Ray, Efficient and cost-effective photoelectrochemical degradation of dyes in wastewater over an exfoliated graphite-MoO_3 nanocomposite electrode. Electrocatalysis 1–9 (2018)
94. T. Ndlovu, O.A. Arotiba, S. Sampath, R.W. Krause, B.B. Mamba, Electrochemical detection and removal of lead in water using poly (propylene imine) modified re-compressed exfoliated graphite electrodes. J. Appl. Electrochem. **41**(12), 1389–1396 (2011)
95. O.M. Ama, Synthesis, characterisation and photoelectrochemical studies of graphite/zinc oxide nanocomposites with the application exfoliated electrodes for the degradation of methylene blue. Int. J. Nano Med. Eng. **2**(8), 145–151 (2017)

96. M. Toyoda, M. Inagaki, Heavy oil sorption using exfoliated graphite: new application of exfoliated graphite to protect heavy oil pollution. Carbon **38**(2), 199–210 (2000)
97. A. Goshadrou, A. Moheb, Continuous fixed bed adsorption of CI Acid Blue 92 by exfoliated graphite: an experimental and modeling study. Desalination **269**(1–3), 170–176 (2011)
98. B. Tryba, A.W. Morawski, M. Inagaki, Application of TiO_2-mounted activated carbon to the removal of phenol from water. Appl. Catal. B **41**(4), 427–433 (2003)
99. K. Yong, Z.L. Wang, W. Yu, Y. Jia, Z.D. Chen, Degradation of methyl orange in artificial wastewater through electrochemical oxidation using exfoliated graphite electrode. New Carbon Mater. **26**(6), 459–464 (2011)
100. D.D.L. Chung, Exfoliation of graphite. J. Mater. Sci. **22**(12), 4190–4198 (1987)
101. K. Parvez, Z.S. Wu, R. Li, X. Liu, R. Graf, X. Feng, K. Mullen, Exfoliation of graphite into graphene in aqueous solutions of inorganic salts. J. Am. Chem. Soc. **136**(16), 6083–6091 (2014)
102. A. Yu, P. Ramesh, M.E. Itkis, E. Bekyarova, R.C. Haddon, Graphite nanoplatelet—epoxy composite thermal interface materials. J. Phys. Chem. C **111**(21), 7565–7569 (2007)
103. G.Z. Kyzas, E.A. Deliyanni, K.A. Matis, Graphene oxide and its application as an adsorbent for wastewater treatment. J. Chem. Technol. Biotechnol. **89**(2), 196–205 (2014)
104. Y. Yang, J. Wang, J. Zhang, J. Liu, X. Yang, H. Zhao, Exfoliated graphite oxide decorated by PDMAEMA chains and polymer particles. Langmuir **25**(19), 11808–11814 (2009)
105. Y. Zhu, S. Murali, W. Cai, X. Li, J.W. Suk, J.R. Potts, R.S. Ruoff, Graphene and graphene oxide: synthesis, properties, and applications. Adv. Mater. **22**(35), 3906–3924 (2010)
106. D.D.L. Chung, A review of exfoliated graphite. J. Mater. Sci. **51**(1), 554–568 (2016)

Chapter 3
The Essence of Electrochemical Measurements on Corrosion Characterization and Electrochemistry Application

Khotso Khoele, Onoyivwe Monday Ama, David Jacobus Delport,
Ikenna Chibuzor Emeji, Peter Ogbemudia Osifo, and Suprakas Sinha Ray

Abstract Electrochemical measurements (EMs) are utilized on corrosion studies and electrochemistry applications. Corrosion is a naturally occurring phenomenon, and that negatively affect functionality of engineering components. To prevent and mitigate it, EMs such as Open circuit potential (OCP), Potentiodynamic Polarization (PDP) and Electrochemical Impedance Spectroscope (EIS) are carried out. Details on all EMs which are carried out to characterize corrosion are included in this study. As twenty-first century industrialization grows, production of inks and dyes for various industrial also grow in direct proportional order. From eventual usage of products manufactured with inks and dyes, different elements become present in wastewater. This phenomenon consists risk to human being, plant and animals. Hence, EMs is necessitated to remove pollutants from the wastewater. On a quite number of EMs, photoelectrochemical technique (PET) is worthwhile on disintegration of havoc in wastewater. PET is carried out within an electrochemical cell (EC). In demonstration of the EMs impact, materials which are used as anodes (photoanodes) within the EC and the engaged organic pollutant simulating solution are thoroughly profiled before and after EMs. All undertakings give necessary information needed on degradation of organic pollutants from the wastewater. Hence, this chapter explains the following important aspects on Electrochemistry application: (i) PET technique

K. Khoele · O. M. Ama · S. S. Ray
DST-CSIR National Center for Nanostructured Materials, Council for Scientific and Industrial Research, Pretoria 0001, South Africa

K. Khoele · D. J. Delport
Department of Chemical, Metallurgical and Materials Engineering, Tshwane University of Technology, Pretoria, South Africa

O. M. Ama (✉) · S. S. Ray
Department of Chemical Science, University of Johannesburg, Doornfontein, 2028, Johannesburg, South Africa
e-mail: onoyivwe4real@gmail.com

O. M. Ama · I. C. Emeji · P. O. Osifo
Department of Chemical Engineering, Vaal University of Technology, Private Mail Bag X021, Vanderbijlpark 1900, South Africa

© Springer Nature Switzerland AG 2020   39
O. M. Ama and S. S. Ray (eds.), *Nanostructured Metal-Oxide Electrode Materials for Water Purification*, Engineering Materials,
https://doi.org/10.1007/978-3-030-43346-8_3

significance and its functionality which make it supersedes other techniques. (ii) A standard EC operation and the constituents elements for successful degradation of wastewater. (iii) Pre and post-measurements techniques which are used as underpinnings demonstrating effectiveness of EMs. (iv) Overall, significance of engaging EMs on wastewater treatment.

Graphical Abstract Corrosion is a tendency of metallic materials returning back to their original states. Therefore; the electrochemistry is a study of electrons migration during metallic substrates application. This shows Corrosion and electrochemistry involve same techniques, but different interpretations.

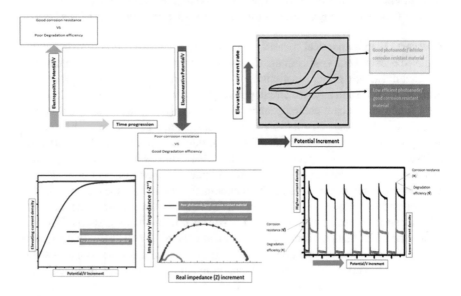

Highlights
In this chapter, the following are included:

- Definition of Corrosion and Electrochemistry with their difference pin-pointed.
- Electrochemical measurements which are utilized under both Corrosion and Electrochemistry, but have opposite interpretations.
- The essence of electrochemical measurements on corrosion characterization and Electrochemistry applications.

Corrosion
Synopses 1

Metals are extracted from mined ores, and they are refined to produce useful products which make great impact to mankind. Alloys are some of the interesting products produced from the mined ores. Nonetheless, during their applications, alloys thermo-dynamically turn to go back to their natural state (ores). The phenomenon pertaining to that process is termed corrosion. Corrosion involves the loss of electrons to the contact area/environment. Eventuality of corrosion encompasses negative chaos such as catastrophic failure which sometimes leads to explosion of structures and death of people. Hence, metals and alloys are studied under their area of applications so as to monitor the potential of corrosion and its kinetics.

Graphical Abstract 1

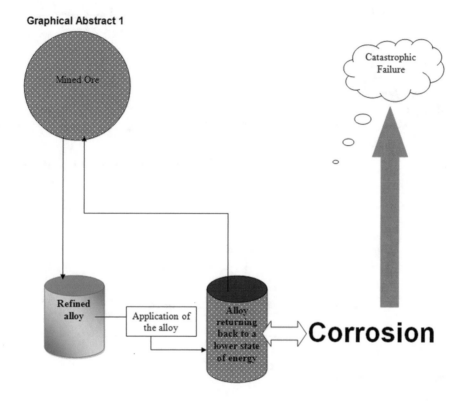

Electrochemistry
Synopses 2

In spontaneous applications of metallic substrates throughout different environments, particularly where electricity is involved, there is always transfer of electrons. This phenomenon is termed Electrochemistry. The application of the Electrochemistry has been proved to solve some problematic phenomena, and removal of organic pollutants from the wastewater is one of the major ones. The recent technique which is used on wastewater degradation is photoelectrochemical oxidation.

Photodegradation Setup

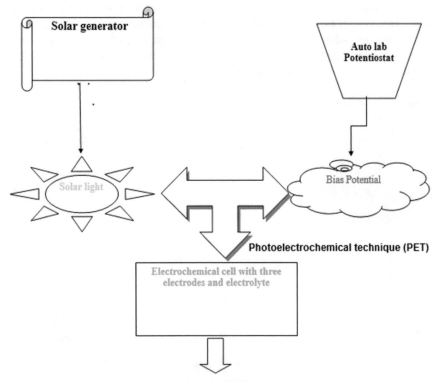

Less harmful organic pollutant (H₂O, CO₂, etc.)

3.1 Introduction

Tons of production on metallic substrates (MSs) is a foundation of twenty-first century industrialization plethora. A nature of MSs makes them to conduct electricity during their applications. Furthermore, within the same MS, there is a variance of the potential. So, when there is moist, a complete electrical circuit (EC) forms [1, 2]. An EC is made out of four major components, and they are: anodic side, cathodic side, electrolyte and the potential difference. An existence of the EC on MSs under application is deemed negative due to havoc that comes out of it. A presence of the EC on MSs is termed corrosion [3–10].

Although metallic substrates very susceptible to corrosion on engineering structures (e.g. pipelines) are either avoided or continuously being improved [11–20], they are beneficial on the other application. Metallic substrates susceptible to corrosion emit electrons faster to the electrolyte, and that causes faster corrosion rates [21–30].

On engineering structures, MSs emitting electrons faster lead to a localized corrosion, and eventually a catastrophic failure. However, the same metallic substrates emitting electrons faster and projecting faster corrosion rates are ideal to disintegrate organic pollutants found in wastewater [31–35]. This phenomenon is beneficial in electrochemistry applications.

Organic pollutants (OPs) are causative of elevating toxicity level in wastewater. Presence OPs in wastewater is from the various sources such as dumped and disposed pharmaceutical products, hospital drainage, house hold, etc. [36–45]. From all the disposals, organic pollutants eventually drive their way into the wastewater. In fact, about 15% of organic pollutants are annually lost into the wastewater, and these comprised negative impacts [36, 46–49].

Wastewater becomes toxic due to consisted elements from OPs [50], and this becomes a risk to human life. In addition, an incomplete mineralization of their processes within wastewater cause water pollution, and negatively affects the environment. Furthermore, their mere disposal kills plants and extent water pollution problems [50]. Hence, it is of paramount importance to degrade organic pollutants from the wastewater. For remediation OPs in wastewater, Electrochemistry application through governing techniques is highly effective tool.

Photoelectrochemical Technique (PET) is one the governing techniques within Electrochemistry application and it are carried out on an electrochemical cell (EC) under electrochemical measurements (EMs). On its operation, a solar light is incorporated to a standard EM. Operationally, a bias potential from the EM and a light from the solar generator are applied simultaneously to immobilize conducting material on a photoanode to generate hydroxyl species which are imperative for the degradation of wastewater pollutants [51]. Utilization of EMs has so far played vital role in degradation of wastewater pollutants. Due to this significance, PET as Ems part has been described in detail within this chapter. An operation of standard EC and pertaining components are thoroughly explained. Furthermore, pre and post-measurements techniques which are mandatorily accompanying the utilization of PET and corrosion in EMs are elaborated. Most importantly, essence of using EMs to degrade organic pollutants from the wastewater is clearly stated. Moreover, significance of engaging corrosion studies in EMs is discussed.

3.2   Corrosion

3.2.1   Corrosion Process

Corrosion involves a transfer of electrons from the more negative part of the MS to the more positive one whenever there is a contact with the electrolyte [52–56]. For example, when steel substrate is in contact with water (H_2O), it corrodes. This occurs due to a loss of electrons (Fe^{2+}) to the Water (H_2O^-) as shown in Fig. 3.1. The SS emits electrons and they are accepted within water molecules (hence water becomes

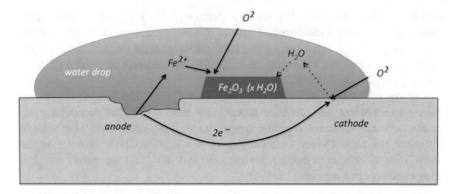

Fig. 3.1 An oxidation formation on the steel substrate due to contact with moisture [57]

positive). An occurrence of the whole process eventually lead to a formation of thick porous layer ($Fe_2O_3(xH_2O)$) as shown in Fig. 3.1.

3.3 Electrochemistry

Electrochemistry consists of a scientific application which deals with an interaction between emitted electrons and chemical species within an aqueous solution. Similar to corrosion cell, Electrochemistry is composed of three electrodes: working, reference and counter electrodes.

Within the phenomenon of Electrochemistry, researchers profiled interaction of electricity with chemical species. This is particularly interesting on anodic side of the EC. On the anodic side, electrons are being emitted from the photoanode to the engaged electrolyte, and that is the process which leads to degradation of organic pollutants in wastewater. So, the faster the electrons being emitted from the photoanode, the better the degradation process. The electrochemical technique that is presently relied upon is PET.

PET is a combination of Electrochemical Oxidation and Photolysis technique. From electrochemical oxidation side, PET is composed of all components the normal EC has. For Photolysis addition, a solar source is added to provide illumination that is needed as can be seen in Fig. 3.2. A synergy operation of PET significantly produces imperatively needed hydroxyl species. In fact, electrons are generated from the photoelectrode and react with released oxygen to develop oxygen reactive radicals which react with protons to generate H_2O^- degrade wastewater pollutants [58–64].

In contrast to corrosion, a working electrode is not always a metal or an alloy. The working electrode is fabricated from the composite, and it is compressed to a pellet form. Furthermore, the working electrode should absorb and transmit light (photo active) [65–69]. Moreover, the electrolyte is composed of the synthetic dye

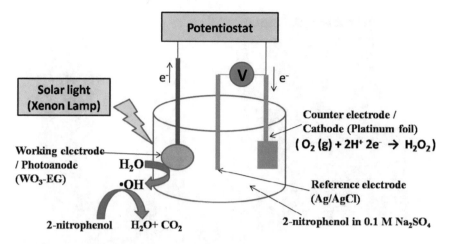

Fig. 3.2   PET in application on degradation of 2-nitrophenol [65]

and some additional materials. So, the distinction between corrosion measurements and Electrochemistry application is the working electrode and the electrolyte.

3.4   Electrochemical Measurements

3.4.1   Pre-measurements and Post Characterization Techniques

On corrosion measurements, pre-measurements and post-characterization techniques are profiling impact of alloying, coating and cathodic/anodic protection relative to the bare MSs. Some of the mostly utilized characterization techniques encompass: scanning electron microscope (SEM), Energy Dispersive X-Ray (EDX), X-Ray Diffraction (XRD), Raman Spectroscope and X-ray spectroscope (XPS). Then, measurements are carried out through AUTO LAB potentiometer equipment shown in Fig. 3.3.

3.4.2   Electrochemical Cell

3.4.2.1   Corrosion Cell Components

Reference Electrode

Reference electrode (RE) measures potential values from working electrode when the voltage is applied on it. This is enabled by the conductive elements within the

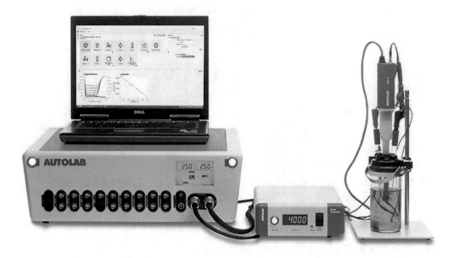

Fig. 3.3 AUTO LAB equipment used for both corrosion measurements and electrochemistry techniques

RE such as Ag/AgCl which is predominantly used for laboratory measurements or Cu/CuSO$_4$ used for empirical applications.

Counter Electrode

As the ohms law states, whenever the potential is applied, there is always a resultant current. The current measurement within the EC is carried out by the use of the counter electrode. The counter electrode (CE) is predominantly a graphite or platinum wire.

Working Electrode

Working electrode (WE) is the most important electrode within the EC as it is the one that necessitates all the other components of the EC and EMs generally. This is where the phenomenon of corrosion is thoroughly profiled, and the WE represents an alloy or metal that is empirically used.

Electrolyte

An electrolyte is a conductive solution which is used to simulate the real-life environment MSs are applied to. For example, buried steel pipelines are exposed to soil externally, and liquid such as oil or water internally [70–79]. The electrolyte is cate-gorically a causative of corrosion, and hence it is engaged on corrosion measurements as to observe corrosion behavior of the MSs in contact with it.

3.4.3   Corrosion Measurements Governing Techniques in EMS

3.4.3.1   Open Circuit Potential (OCP)

Before an application of any corrosion technique, OCP is applied on EMs. This is done with no current flowing within the EC setup. As mentioned earlier that corrosion takes place whenever there is existence of the four components: cathodic reaction, anodic reaction, electrolyte and the potential difference. The OCP is the point of equilibrium during the con-occurrence of the four mentioned components. So, for the corrosion to take place, the cathodic and anodic reactions take place simultaneously at the equal rate. Hence the OCP is the point of intersection between anodic and cathodic reactions as can be seen on the Evans diagram in Fig. 3.4. Clearly, OCP is denoted vertically by E_{corr} at the potential side and by I_{corr} horizontally to the current density side.

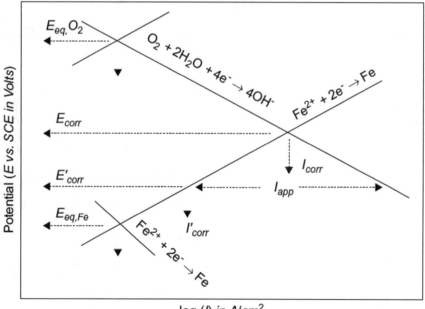

Fig. 3.4   Evans diagram showing OCP and polarization of LCSP from OCP [80]

3.4.3.2 Potentiodynamic Polarization (PDP)

Utilization of PDP measurements on EMs is wholly for examination of uniform corrosion on MSs. Deductions is obtained from cathodic and anodic curves drawn. In particular, if the anodic curve keeps extending to right-hand side with an increase of current density, that shows continuing corrosion process. However, if the anodic curve at one point becomes constant, at the certain current density, there is an occurrence of passivation on the MS. Clear distinction can be seen from Fig. 3.5 which shows corroding alloy without the formation of intrinsic passivation layer, and Fig. 3.6 on demonstration of noble alloy which formed an intrinsic oxide layer when is in contact with the electrolyte.

Fig. 3.5 PDP for porous Magnesium and Magnesium with Germanium different concentrations [81]

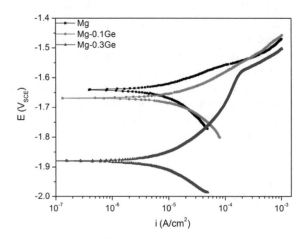

Fig. 3.6 Bare SAC 105 and addition of different concentration of Aluminum [82]

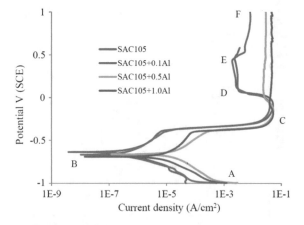

3.4.3.3 Cyclic PDP

MSs are different in terms of their natural way of corroding. In fact, the oxide layer which forms when MSs corrode is either a thin oxide layer or porous dense oxide layer. For instance, earth-rare metals and their alloys form an interesting thin layer which retard extent of corrosion on their surfaces [81–84]. Nonetheless, general metals such as Fe, Mg and their alloys such as low-carbon steel form a porous layer which is very susceptible to corrosion [85–89]. Hence, MSS have different corrosion behaviour. To characterize localized corrosion behaviour on MSs, Cyclic PDP is used for two observations. First, it is used to characterize localized corrosion behaviour on MSs under different applications. The main reason for examination is to probe a point where pitting initiate and its propagation behaviour. Second, as MSs and alloys are coated or substituted with newly selected materials, Cyclic PDP characterizes distinction on the reduction of the hysteresis loop of the modified material in relative to the standard. Figure 3.7 shows the cyclic PDP curves on 17-4PH stainless steel at low and high sulphur content in 3.5% NaCl solution, while Fig. 3.8 shows cyclic PDP curve on three stainless steel under different volumes of NaCl solution. From both Figs, it can be seen that reverse scans proceed behind first scans. Furthermore, it can be noticed that some reverse scans pass above OCP points while others go below. The ones passing above show that there is a likelihood of pitting initiation, while the ones passing underneath reveal that there is both tendency to pit initiation and propagation. Contrarily, opposite curves are demonstrated in Fig. 3.9. The demonstrated cyclic PDP curves are for pure Titanium and titanium alloy ($Ti_{20}Nb_{10}Zr_5Ta$) under Ringer solution. It can be observed on all curves that the reverse scans are all facing forward, and that shows that there is neither tendency for pitting initiation nor propagation. Only the best can be picked in terms of wider forward reverse scan.

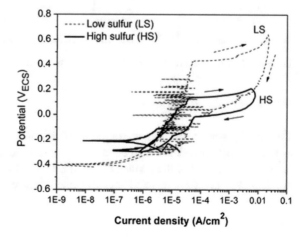

Fig. 3.7 17-4PH stainless steel Cyclic PDP at low and high sulfur content in 3.5% NaCl solution [90]

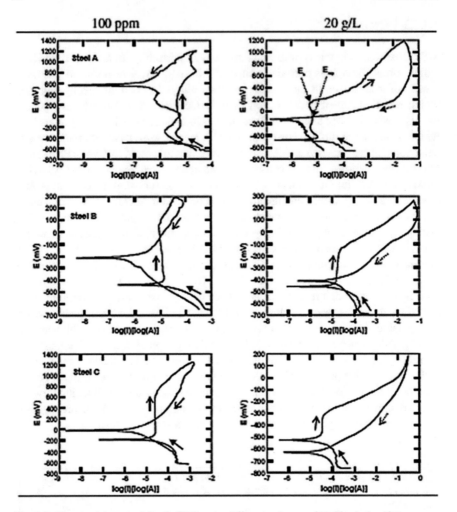

Fig. 3.8 Three stainless steel Cyclic PDP under different volumes of NaCl solution [91]

3.4.3.4 Chronoamperometry

Chronoamperometry technique examines relationship between current and potential as the potential is varied within certain amount of time. A variation of the potential is used to characterize the surface of the WE within the electrolyte. Measurements are displayed in a form of fixed potential against current and time as can be seen in Fig. 3.10.

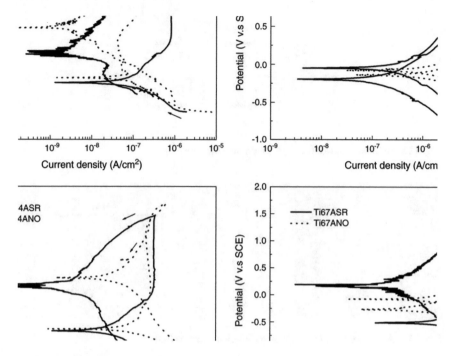

Fig. 3.9 Cyclic PDP scans on different Titanium alloys [92]

Fig. 3.10 Current-time
curves for API steel
electrodes from 1 to 2 h,
respectively [93]

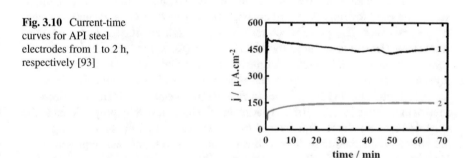

3.4.3.5   Electrochemical Impedance Spectroscopy (EIS)

EIS measures and models properties of the bare and/or coated MSs Over a series
of frequency (low, middle and high) at OCP and under applied potentials [94]. On
characterization of the coated surfaces, EIS reveals the information of great importance in terms of the dielectric properties. In particular, the state organic coatings
are presently probed, with the use of an equivalent electrical circuit (EEC), is in two
categories: semi-perfect (initial breakdown) EEC and full damaged coating ECC.

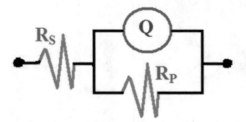

Fig. 3.11 An initial breakdown of coating/semi perfect coating [93]

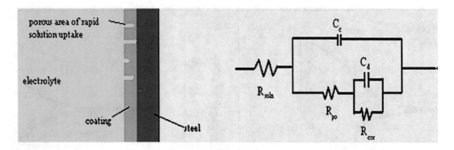

Fig. 3.12 A damaged coating at the metal/solution interface [93]

Figure 3.11 shows the EEC of the semi-perfect coating, while Fig. 3.12 demonstrates the EEC for fully damaged coating. The engaged EEC parameters are electrolyte resistance (R_s), polarization resistance (R_p), solution resistance (R_{soln}), constant phase element (Q), charge transfer resistance (Rct), film capacitance of double layer capacitance at the interface of metal and electrolyte (C_c) and substrate double layer capacitance (C_d). These parameters are normally obtained from Nyquist spectra and are fitted to the EEC in order to characterize whether corrosion reactions are mild or aggressive [83, 95–105]. In particular, the R_{ct} is inversely proportional to the corrosion rate. So, the R_{ct} from Nyquist spectra modeled by EEC shown in Fig. 3.11 are normally obtained to be higher in values in relative to the ones represented by Nyquist diagrams fitted with EEC shown in Fig. 3.12. Under the conditions represented by EEC shown in Fig. 3.12, the coating is aggressively damaged, and that leads to higher corrosion rates. Typically, the Nyquist diagrams with wider diameters are represented by the EEC shown in Fig. 3.11, while the most compressed Nyquist diagrams correspond to the EEC shown in Fig. 3.12.

3.4.3.6 Cyclic Voltammetry (CV)

As corrosion rates are equivalency of current rates, CV is applied to reveal corrosion rates per certain area (current density). CV curves assist in terms of revealing how fast the corrosion rates are. The higher the current density within engaged potential

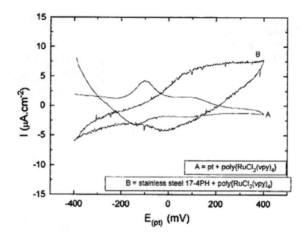

Fig. 3.13 CV measurements between two substrates A and B [106]

range, the faster the corrosion rates. Figure 3.13 shows a comparison between two substrates A and B. Based on analysis deduced from the curves, it can be seen that substrate A exhibits lower current densities, and that shows lower corrosion rates in relative to the substrate B in elevation of the potential.

3.4.4   Electrochemistry Application

3.4.4.1   Pre-electrochemistry Application

Preparation of Powders to Form a Pellet and Composite

Powders in a form of Nano or micro-particles are normally bought in a form of reagents, and then mixed together with appropriate solutions such as sulfuric acid (H_2SO_4), etc., to form an intended composite. A well-mixed and blended composite is taken to a hydraulic press under high temperature, and compressed to a pellet of about 1.3 cm diameters [35].

Working Electrode

The WE under electrochemistry application is made from the fabricated pellet. There are generally three steps which are carried out to form the WE (photoelectrode). First, a one end of a copper (Cu) wire is inserted and passed out through a glass tube with an opening at both ends and coiled onto the fabricated nanocomposite pellet. Secondly, conduction between the Cu wire and the pellet is created with a use of conductive silver paste. Thirdly, a proper close-up on the glass tube is done by covering up with an epoxy resin at all ends. That is done to prohibit electrolyte entrance into the tube.

Fig. 3.14 Curves were
drawn from different
temperature investigations
[111]

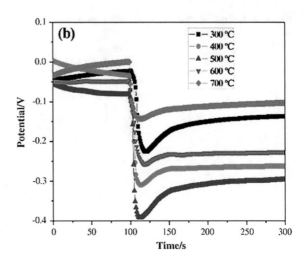

Electrolyte

Electrolyte of the EC for Electrochemistry application is composed of synthetic
dye solution such as potassium ferricyanide, ferrocyanide, KCl, etc. Furthermore,
an additional solution such as Na_2SO_4, NaOH and HCl are normally added as a
supporting solution to make the electrolyte more conductive [35, 47, 48, 107–110],
etc.

3.4.5 Electrochemistry Governing Techniques on EMs

3.4.5.1 OCP

Unlike on corrosion studies, OCP measurements under Electrochemistry application are specifically on degradation kinetics. The photoelectrode showing more
electronegative potential/voltage (E/V) is normally the most efficient one. The perfect example of OCP engagement under Electrochemistry application is shown in
Fig. 3.14. Curves were drawn from different temperature investigations, and 550 °C
was noted to be the most efficient [111].

3.4.5.2 Cyclic Voltammetry

CV is used mainly to compare current and peak-to-peak separation (Faradaic current)
from engaged materials on degradation processes. Some of the significance output
of the CV application on EMs is shown in Fig. 3.15. From all the curves, it can be
seen that illuminated conditions show relatively higher peak on resultant current.

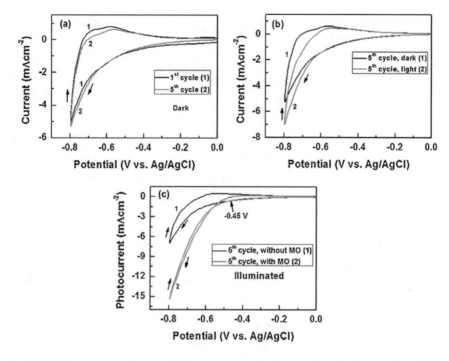

Fig. 3.15 CV measurements on two different cycles within dark and illumination conditions [112]

That shows faster degradation on illuminated electrodes, and that is ideal efficiency Electrochemistry application intents to achieve on removal of organic pollutants from the wastewater treatment.

3.4.5.3 Linear Sweep Voltammetry (LSV)

LSV is specifically engaged in EMs to examine the behaviour of engaged electrodes in terms of degradation efficiency on wastewater pollutants disintegration under either illumination or darkness so as to characterize degradation efficiency. The LSV application can be seen in Fig. 3.16. From the analysis, it was observed that WO_3-EG under illumination exhibit higher current rates, and that proved better degradation efficiency.

3.4.5.4 Square Wave

For further validation of the other EMs, Square Wave is applied to produce a Photo-current response (PR) curves. PR curves assist to reveal and confirm whether charge

Fig. 3.16 LSV on EG and
WO$_3$ under light and dark
application [65]

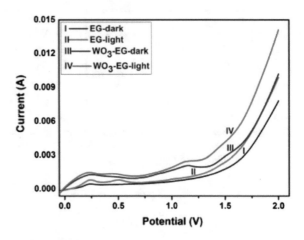

Fig. 3.17 Photocurrent
response curves from the
PET application [65]

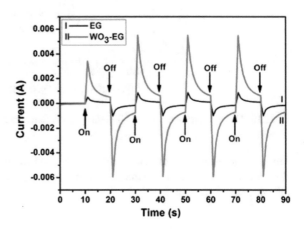

separation occurred on the engaged electrodes. In particular, efficiency of the relatively modified electrode to standard electrode is what is being clarified the most. Figure 3.17 shows the PR curves between EG as the standard electrode and WO$_3$-EG as the modified electrode. From the analysis, it was discovered that WO$_3$-EG exhibit higher peaks of current within recorded 90 min as can be seen in Fig. 3.17. This is attributed to a better degradation efficiency offered by WO$_3$-EG [65].

3.5   Summary and Future Scope of Electrochemical Measurements

Electrochemical measurements are chief methodological tools on corrosion studies and Electrochemistry application.

Electrochemistry is a scientific phenomenon which occurs when spontaneous/non-spontaneous flow of electrons interacts with chemical species from an aqueous solution.

On continuing diminishing of portable water, Electrochemistry application through PET is good for the removal of organic pollutants from the wastewater.

Corrosion is a natural phenomenon occurring to all metallic substrates.

Studies of corrosion on metallic substrates are imperative on endless avoidance of catastrophic failure of engineering components during the applications.

AUTOLAB Potentiometer is central equipment which is used for electrochemical measurements for corrosion studies and electrochemical applications.

The electrochemical cell is the same for both corrosion studies and electrochemical applications, and it is composed of three electrodes and electrolyte:

- **Reference electrode (RE)** is primarily used to measure the potential on working electrode when the voltage is applied on it. This is enabled by the conductive elements within the RE such as Ag/AgCl or Cu/CuSO$_4$.
- **Counter electrode (CE)** is used to measure the resultant current on working electrode when the voltage is applied on it.
- **Working electrode (WE)** is the simulator of empirical substance under the study. In corrosion, working electrode represents an alloy or metal that is empirically used for a particular industrial application, while in Electrochemistry application WE is the fabricated photo composite.
- **Electrolyte** is a conductive solution which when combines with anodic and cathodic reactions, in existence of potential difference along, forms an electrochemical cell.

Photoelectrochemical technique (PET) is an Electrochemistry tool which is presently used in wastewater degradation. This technique uses both solar light and electrochemical oxidation to degrade organic pollutants from the wastewater.

Most of the governing techniques under both corrosion measurements and Electrochemistry are kind of reverse of each other, and the mostly utilized measurements under both corrosion studies and electrochemistry application are: Open circuit potential (OCP), Potentiodynamic Polarization (PDP), Cyclic PDP, Chronoamperometry, Electrochemical impedance spectroscope (EIS), Cyclic Voltammetry (CV) and Linear Sweep Voltammetry (LSV).

3.6   Conclusions and Future Scope

3.6.1   Conclusions

Critical Electrochemical Measurements utilized in corrosion studies and Electrochemistry applications have been discussed in detailed within this chapter.

On functionalization, modification, alloying, coating, etc., of diversity materials under different applications, Electrochemical measurements and governing characterization tools reveal whether the pin-pointed aim has successfully been achieved or not. There are about five electrochemical measurements which are used in both corrosion studies and Electrochemistry application.

The Electrochemistry application in this study has been focused on degradation of wastewater pollutants using Photoelectrochemical technique (PET).

PET is applied on fabricated photoelectrodes to produce oxygen and some other reactive radicals which react with protons to generate hydroxyl ions (OH^-) which degrade wastewater pollutants. The degraded products are consisting of H_2O, CO_2 and inorganic products such as NO_3^-, etc.

Some of the most important aspects which make analysis of the electrochemical measurements to differ between corrosion studies and Electrochemistry application are the following:

- The more negative Open Circuit Potential (OCP) values, under corrosion measurements are, show tendency of higher corrosion potential which could be causative of eventual catastrophic failure. However, more negative OCP values under Electrochemistry application shows thrilling impact which assist in generation of the more OH^- ideal for efficient degradation processes.
- On Cyclic Voltammograms (CV), the emerging of peaks which go along with exhibition of higher current density shows fast occurrence of corrosion reactions on the surface of the working electrode, while it shows sought-after characteristics from engaged photoelectrode on utilization of PET.
- Under Electrochemical impedance spectroscope (EIS) measurements, the most compressed semi-circle from Nyquist diagrams correspond to material which has poor corrosion resistance properties. However, the same compressed semi-circle under Electrochemistry shows photoelectrode with the most degradation efficiency on degradation of organic pollutants from the wastewater.
- Overall, Electrochemical measurements are vice-verse of each other between corrosion studies and Electrochemistry application.

3.6.2 Future Scope

Corrosion phenomena affect all engineering components, and it is thermodynamically inevitable. To date, ways of mitigating corrosion on Engineering tools encompass three methods: coating, alloying, material selection and design.

On continual improvement, some of the critical areas under applications are on coating of steel pipelines with organic coatings, Surface engineering of alloys, alloying using inert materials and improvement of designs. Within the scope, different complications necessitating betterment have been observed:

First, organic coatings which are used on buried steel pipelines fail due to coatings' disbondment. This is deemed to be emanating from soil heterogeneity. Hence, empirical and laboratory characterization techniques should be improved to give exact occurrences soil heterogeneity impose on buried steel pipelines.

Second, irrespective of using inert materials on alloying of new materials, surfaces are still prone to failure due to wear and corrosion. Surface engineering techniques such as laser deposition, metallic coating, etc., found to be inroad-making, and therefore should further be improved.

Third, in medical and dental applications, alloying using earth-rare metals such zirconium, titanium, platinum, gold, etc., proved to yield products which are relatively corrosion resistant and have superior mechanical properties. However, earth-rare metals are diminishing, and thus alternative cheap, bio-compatible and readily available materials that could have equivalent corrosion resistance and better mechanical properties equivalent to that of earth-rare metals should continuously be investigated.

Fourthly, useful engineering products such as pumps and water treatment plants continue to fail at the intricate part of their designs. Therefore, new designs, particularly innovative at elbows and jaw-ends, should be put into place.

Electrochemistry application on wastewater treatment using various photocatalysists is indisputably profound. Presently, photoelectrochemical technique in combination with synergistic photocatalysists has been reported useful. To improve efficiency of the degradation, further combination of photoanodes in a form of binary, ternary and quaternary should be engaged.

References

1. A. Abdullah, N. Yahaya, N. Md Noor, R. Mohd Rasol, Microbial corrosion of API 5L X-70 carbon steel by ATCC 7757 and consortium of sulfate-reducing bacteria. J. Chem. (2014)
2. O. Abootalebi, A. Kermanpur, M.R. Shishesaz, M.A. Golozar, Optimizing the electrode position in sacrificial anode cathodic protection systems using boundary element method. Corros. Sci. **52**(3), 678–687 (2010)
3. M.A. Abu-Dalo, N.A. Al-Rawashdeh, A. Ababneh, Evaluating the performance of sulfonated Kraft lignin agent as corrosion inhibitor for iron-based materials in water distribution systems. Desalination **313**, 105–114 (2013)
4. S.A. Ajeel, G.A. Ali, Variable conditions effect on polarization parameters of impressed current cathodic protection of low carbon steel pipes. Eng. Technol. J. **26**(6), 636–647 (2008)
5. E. Akbarzadeh, M.M. Ibrahim, A.A. Rahim, Corrosion inhibition of mild steel in near neutral solution by kraft and soda lignins extracted from oil palm empty fruit bunch. Int. J. Electrochem. Sci. **6**(11), 5396–5416 (2011)
6. I.A. Akpan, N.O. Offiong, Electrochemical study of the corrosion inhibition of mild steel in hydrochloric acid by amlodipine drug. Int. J. Chem. Mater. Res. **2**(3), 23–29 (2014)
7. J.L. Alamilla, M.A. Espinosa-Medina, E. Sosa, Modelling steel corrosion damage in soil environment. Corros. Sci. **51**(11), 2628–2638 (2009)
8. J.L. Alamilla, E. Sosa, Stochastic modelling of corrosion damage propagation in active sites from field inspection data. Corros. Sci. **50**(7), 1811–1819 (2008)
9. E. Alkhateeb, R. Ali, N. Popovska-Leipertz, S. Virtanen, Long-term corrosion study of low carbon steel coated with titanium boronitride in simulated soil solution. Electrochim. Acta **76**, 312–319 (2012)
10. I. Andijani, S. Turgoose, Studies on corrosion of carbon steel in deaerated saline solutions in presence of scale inhibitor. Desalination **123**(2–3), 223–231 (1999)

11. A.A. Atshan, B.O. Hasan, M.H. Ali, Effect of anode type and position on the cathodic protection of carbon steel in sea water. Int. J. Current Eng. Technol. **3**(5), 2017–2024 (2013)
12. M. Barbalat, L. Lanarde, D. Caron, M. Meyer, J. Vittonato, F. Castillon et al., Electrochemical study of the corrosion rate of carbon steel in soil: evolution with time and determination of residual corrosion rates under cathodic protection. Corros. Sci. **55**, 246–253 (2012)
13. T. Breton, J.C. Sanchez-Gheno, J.L. Alamilla, J. Alvarez-Ramirez, Identification of failure type in corroded pipelines: a Bayesian probabilistic approach. J. Hazard. Mater. **179**(1–3), 628–634 (2010)
14. F. Caleyo, J.C. Velázquez, A. Valor, J.M. Hallen, Probability distribution of pitting corrosion depth and rate in underground pipelines: a Monte Carlo study. Corros. Sci. **51**(9), 1925–1934 (2009)
15. A. Cervantes-Tobón, J.G. Godínez-Salcedo, J.L. Gonzalez-Velazquez, M. Díaz-Cruz, Corrosion rates of API 5L X-52 and X-65 steels in synthetic brines and brines with H_2S as a function of rate in a rotating cylinder electrode. Int. J. Electrochem. Sci. **9**, 2454–2469 (2014)
16. T. Charng, F. Lansing, Review of corrosion causes and corrosion control in a technical facility. *NASA Technical Report, TDA Progress Report* (1982), pp. 42–69
17. X. Chen, G. Wang, F. Gao, Y. Wang, C. He, Effects of sulphate-reducing bacteria on crevice corrosion in X70 pipeline steel under disbonded coatings. Corros. Sci. **101**, 1–11 (2015)
18. Z. Chik, T. Islam, Study of chemical effects on soil compaction characterizations through electrical conductivity. Int. J. Electrochem. Sci. **6**, 6733–6740 (2011)
19. C. Christodoulou, G. Glass, J. Webb, S. Austin, C. Goodier, Assessing the long term benefits of impressed current cathodic protection. Corros. Sci. **52**(8), 2671–2679 (2010)
20. I.S. Cole, D. Marney, The science of pipe corrosion: a review of the literature on the corrosion of ferrous metals in soils. Corros. Sci. **56**, 5–16 (2012)
21. D.K. Kim, S. Muralidharan, T.H. Ha, J.H. Bae, Y.C. Ha, H.G. Lee, J.D. Scantlebury, Electrochemical studies on the alternating current corrosion of mild steel under cathodic protection condition in marine environments. Electrochim. Acta **51**(25), 5259–5267 (2006)
22. F. Delaunois, F. Tosar, V. Vitry, Corrosion behaviour and biocorrosion of galvanized steel water distribution systems. Bioelectrochemistry **97**, 110–119 (2014)
23. E.O. Eltai, J.D. Scantlebury, E.V. Koroleva, The effects of different ionic migration on the performance of intact unpigmented epoxy coated mild steel under cathodic protection. Prog. Org. Coat. **75**(1–2), 79–85 (2012)
24. A.M. El-Shamy, M.F. Shehata, A.I.M. Ismail, Effect of moisture contents of bentonitic clay on the corrosion behavior of steel pipelines. Appl. Clay Sci. **114**, 461–466 (2015)
25. B.M. Fernández-Pérez, J.A. González-Guzmán, S. González, R.M. Souto, Electrochemical impedance spectroscopy investigation of the corrosion resistance of a waterborne acrylic coating containing active electrochemical pigments for the protection of carbon steel. Int. J. Electrochem. Sci. **9**(4), 2067 (2014)
26. E. Gamboaa, N. Coniglioa, R. Kurjia, G. Callar, Hydrothermal ageing of X65 steel specimens coated with 100% solids epoxy. Prog. Org. Coat. **76**, 1505–1510 (2013)
27. I.M. Gadala, M.A. Wahab, A. Alfantazi, Numerical simulations of soil physicochemistry and aeration influences on the external corrosion and cathodic protection design of buried pipeline steels. Mater. Des. **97**, 287–299 (2016)
28. I.M. Gadala, A. Alfantazi, Electrochemical behavior of API-X100 pipeline steel in NS₄, near-neutral, and mildly alkaline pH simulated soil solutions. Corros. Sci. **82**, 45–57 (2014)
29. N. Guermazi, K. Elleuch, H.F. Ayedi, V. Fridrici, P. Kapsa, Tribological behaviour of pipe coating in dry sliding contact with steel. Mater. Des. **30**(8), 3094–3104 (2009)
30. I. Gurrappa, Cathodic protection of cooling water systems and selection of appropriate materials. J. Mater. Process. Technol. **166**, 256–267 (2005)
31. P. Li, G. Zhao, K. Zhao, J. Gao, T. Wu, An efficient and energy saving approach to photocatalytic degradation of opaque high-chroma methylene blue wastewater by electrocatalytic pre-oxidation. Dyes Pigm. **92**(3), 923–928 (2012)
32. W.K. Jo, R.J. Tayade, Recent developments in photocatalytic dye degradation upon irradiation with energy-efficient light emitting diodes. Chin. J. Catal. **35**(11), 1781–1792 (2014)

33. N. Chaukura, W. Gwenzi, N. Tavengwa, M.M. Manyuchi, Biosorbents for the removal of synthetic organics and emerging pollutants: opportunities and challenges for developing countries. Environ. Dev. **19**, 84–89 (2016)
34. N. Chaukura, B.B. Mamba, S.B. Mishra, Porous materials for the sorption of emerging organic pollutants from aqueous systems: the case for conjugated microporous polymers. J. Water Process Eng. **16**, 223–232 (2017)
35. J.T. Jasper, O.S. Shafaat, M.R. Hoffmann, Electrochemical transformation of trace organic contaminants in latrine wastewater. Environ. Sci. Technol. **50**(18), 10198–10208 (2016)
36. P. Nigam, I.M. Banat, D. Singh, R. Marchant, Microbial process for the decolorization of textile effluent containing azo, diazo and reactive dyes. Process Biochem. **31**(5), 435–442 (1996)
37. A. Mirzaei, Z. Chen, F. Haghighat, L. Yerushalmi, Removal of pharmaceuticals from water by homo/heterogonous Fenton-type processes—a review. Chemosphere **174**, 665–688 (2017)
38. M.J. Ahmed, B.H. Hameed, Removal of emerging pharmaceutical contaminants by adsorption in a fixed-bed column: a review. Ecotoxicol. Environ. Saf. **149**, 257–266 (2018)
39. A. El-Ghenymy, F. Centellas, J.A. Garrido, R.M. Rodríguez, I. Sirés, P.L. Cabot, E. Brillas, Decolorization and mineralization of Orange G azo dye solutions by anodic oxidation with a boron-doped diamond anode in divided and undivided tank reactors. Electrochim. Acta **130**, 568–576 (2014)
40. X. Florenza, S. Garcia-Segura, F. Centellas, E. Brillas, Comparative electrochemical degradation of salicylic and aminosalicylic acids: influence of functional groups on decay kinetics and mineralization. Chemosphere **154**, 171–178 (2016)
41. H.H. Ngo, W. Guo, J. Zhang, S. Liang, C. Ton-That, X. Zhang, Typical low cost biosorbents for adsorptive removal of specific organic pollutants from water. Biores. Technol. **182**, 353–363 (2015)
42. E. do Vale-Júnior, S. Dosta, I.G. Cano, J.M. Guilemany, S. Garcia-Segura, C.A. Martínez-Huitle, Acid blue 29 decolorization and mineralization by anodic oxidation with a cold gas spray synthesized Sn–Cu–Sb alloy anode. Chemosphere **148**, 47–54 (2016)
43. C.P. Silva, G. Jaria, M. Otero, V.I. Esteves, V. Calisto, Waste-based alternative adsorbents for the remediation of pharmaceutical contaminated waters: has a step forward already been taken? Biores. Technol. **250**, 888–901 (2018)
44. D.W. Li, W.L. Zhai, Y.T. Li, Y.T. Long, Recent progress in surface enhanced Raman spectroscopy for the detection of environmental pollutants. Microchim. Acta **181**(1–2), 23–43 (2014)
45. S.F. Hasany, I. Ahmed, J. Rajan, A. Rehman, Systematic review of the preparation techniques of iron oxide magnetic nanoparticles. Nanosci. Nanotechnol. **2**(6), 148–158 (2012)
46. A.M.S. Solano, C.A. Martínez-Huitle, S. Garcia-Segura, A. El-Ghenymy, E. Brillas, Application of electrochemical advanced oxidation processes with a boron-doped diamond anode to degrade acidic solutions of Reactive Blue 15 (Turqueoise Blue) dye. Electrochim. Acta **197**, 210–220 (2016)
47. X.L. He, C. Song, Y.Y. Li, N. Wang, L. Xu, X. Han, D.S. Wei, Efficient degradation of azo dyes by a newly isolated fungus Trichoderma tomentosum under non-sterile conditions. Ecotoxicol. Environ. Saf. **150**, 232–239 (2018)
48. S. Garcia-Segura, E. Brillas, Applied photoelectrocatalysis on the degradation of organic pollutants in wastewaters. J. Photochem. Photobiol., C **31**, 1–35 (2017)
49. S. Natarajan, H.C. Bajaj, R.J. Tayade, Recent advances based on the synergetic effect of adsorption for removal of dyes from waste water using photocatalytic process. J. Environ. Sci. **65**, 201–222 (2018)
50. R.M. Fernández-Domene, R. Sánchez-Tovar, B. Lucas-granados, M.J. Munoz-Portero, J. García-Antón, Elimination of pesticide atrazine by photoelectrocatalysis using a photoanode based on WO_3 nanosheets. Chem. Eng. J. **350**, 1114–1124 (2018)
51. S.E. Sanni, A.P. Ewetade, M.E. Emetere, O. Agboola, E. Okoro, S.J. Olorunshola, T.S. Olugbenga, Enhancing the inhibition potential of sodium tungstate towards mitigating the corrosive effect of Acidithiobacillus thiooxidan on X-52 carbon steel. Mater. Today Commun. **19**, 238–251 (2019)

52. Y. Cui, T. Shen, Q. Ding, Study on the Influence of AC Stray Current on X80 Steel under Stripped Coating by Electrochemical Method. Int. J. Corrosion (2019)
53. M. Kiani Khouzani, A. Bahrami, A. Hosseini-Abari, M. Khandouzi, P. Taheri, Microbiologically influenced corrosion of a pipeline in a petrochemical plant. Metals **9**(4), 459 (2019)
54. W.A. Byrd, Below grade coated direct imbedded steel pole corrosion failures with solutions. IEEE Power Energy Technol. Syst. J. **6**(1), 41–46 (2019)
55. S. Papavinasam, *Corrosion Control in the Oil and Gas Industry* (Elsevier, Amsterdam, 2013)
56. R. Jia, T. Unsal, D. Xu, Y. Lekbach, T. Gu, Microbiologically influenced corrosion and current mitigation strategies: a state of the art review. Int. Biodeterior. Biodegrad. **137**, 42–58 (2019)
57. D.H. Spennemann, Why do corroded corrugated iron roofs have a striped appearance? (2015)
58. O.M. Ama, K. Khoele, W.W. Anku, S.S. Ray, Photoelectrochemical degradation of 4-nitrophenol using CuO–ZnO/exfoliated graphite nanocomposite electrode. Int. J. Electrochem. Sci. **14**(3), 2893–2905 (2019)
59. X. Fan, Y. Zhou, G. Zhang, T. Liu, W. Dong, In situ photoelectrochemical activation of sulfite by MoS_2 photoanode for enhanced removal of ammonium nitrogen from wastewater. Appl. Catal. B **244**, 396–406 (2019)
60. X. Wang, Q. Wu, H. Ma, C. Ma, Z. Yu, Y. Fu, X. Dong, Fabrication of PbO_2 tipped Co_3O_4 nanowires for efficient photoelectrochemical decolorization of dye (reactive brilliant blue KN-R) wastewater. Sol. Energy Mater. Sol. Cells **191**, 381–388 (2019)
61. R. Matarrese, M. Mascia, A. Vacca, L. Mais, E.M. Usai, M. Ghidelli et al., Integrated Au/TiO_2 nanostructured photoanodes for photoelectrochemical organics degradation. Catalysts **9**(4), 340 (2019)
62. C.F. Liu, C.P. Huang, C.C. Hu, C. Huang, A dual TiO_2/Ti-stainless steel anode for the degradation of orange G in a coupling photoelectrochemical and photo-electro-Fenton system. Sci. Total Environ. **659**, 221–229 (2019)
63. C. Zhang, G. Ren, W. Wang, X. Yu, F. Yu, Q. Zhang, M. Zhou, A new type of continuous-flow heterogeneous electro-fenton reactor for tartrazine degradation. Sep. Purif. Technol. **208**, 76–82 (2019)
64. X. Li, J. Bai, J. Li, Y. Zhang, Z. Shen, L. Qiao, et al., Efficient TN removal and simultaneous TOC conversion for highly toxic organic amines based on a photoelectrochemical-chlorine radicals process. Catal. Today (2019)
65. E.H. Umukoro, M.G. Peleyeju, J.C. Ngila, O.A. Arotiba, Towards wastewater treatment: photo-assisted electrochemical degradation of 2-nitrophenol and orange II dye at a tungsten trioxide-exfoliated graphite composite electrode. Chem. Eng. J. **317**, 290–301 (2017)
66. Z. Fan, H. Shi, H. Zhao, J. Cai, G. Zhao, Application of carbon aerogel electrosorption for enhanced Bi_2WO_6 photoelectrocatalysis and elimination of trace nonylphenol. Carbon **126**, 279–288 (2018)
67. T. Hong, Z. Liu, X. Zheng, J. Zhang, L. Yan, Efficient photoelectrochemical water splitting over Co_3O_4 and Co_3O_4/Ag composite structure. Appl. Catal. B **202**, 454–459 (2017)
68. K. Barbari, R. Delimi, Z. Benredjem, S. Saaidia, A. Djemel, T. Chouchane et al., Photocatalytically-assisted electrooxidation of herbicide fenuron using a new bifunctional electrode PbO_2/SnO_2-$Sb_2O_3/Ti//Ti/TiO_2$. Chemosphere **203**, 1–10 (2018)
69. Z. Wang, M. Xu, F. Wang, X. Liang, Y. Wei, Y. Hu et al., Preparation and characterization of a novel Ce doped PbO_2 electrode based on NiO modified Ti/TiO_2NTs substrate for the electrocatalytic degradation of phenol wastewater. Electrochim. Acta **247**, 535–547 (2017)
70. Y. Yao, B. Ren, Y. Yang, C. Huang, M. Li, Preparation and electrochemical treatment application of Ce-PbO_2/ZrO_2 composite electrode in the degradation of acridine orange by electrochemical advanced oxidation process. J. Hazard. Mater. **361**, 141–151 (2019)
71. S. Malato, P. Fernández-Ibáñez, M.I. Maldonado, J. Blanco, W. Gernjak, Decontamination and disinfection of water by solar photocatalysis: recent overview and trends. Catal. Today **147**(1), 1–59 (2009)
72. M. Canovi, J. Lucchetti, M. Stravalaci, F. Re, D. Moscatelli, P. Bigini et al., Applications of surface plasmon resonance (SPR) for the characterization of nanoparticles developed for biomedical purposes. Sensors **12**(12), 16420–16432 (2012)

73. O.S. Wolfbeis, An overview of nanoparticles commonly used in fluorescent bioimaging. Chem. Soc. Rev. **44**(14), 4743–4768 (2015)
74. I. Poulios, D. Makri, X. Prohaska, Photocatalytic treatment of olive milling waste water: oxidation of protocatechuic acid. Global Nest: Int. J. **1**(1), 55–62 (1999)
75. D. Pathania, R. Katwal, G. Sharma, M. Naushad, M.R. Khan, H. Ala'a, Novel guar gum/Al_2O_3 nanocomposite as an effective photocatalyst for the degradation of malachite green dye. Int. J. Biol. Macromol. **87**, 366–374 (2016)
76. J. Luo, M. Hepel, Photoelectrochemical degradation of naphthol blue black diazo dye on WO_3 film electrode. Electrochim. Acta **46**(19), 2913–2922 (2001)
77. V.M. Daskalaki, M. Antoniadou, G. Li Puma, D.I. Kondarides, P. Lianos, Solar light-responsive $Pt/CdS/TiO_2$ photocatalysts for hydrogen production and simultaneous degradation of inorganic or organic sacrificial agents in wastewater. Environ. Sci. Technol. **44**(19), 7200–7205 (2010)
78. G. Elango, S.M. Roopan, Efficacy of SnO_2 nanoparticles toward photocatalytic degradation of methylene blue dye. J. Photochem. Photobiol., B **155**, 34–38 (2016)
79. I.M. Szilágyi, B. Fórizs, O. Rosseler, Á. Szegedi, P. Németh, P. Király et al., WO_3 photocatalysts: influence of structure and composition. J. Catal. **294**, 119–127 (2012)
80. N. Popov Branko, P. Kumaraguru Swaminatha, *25-Cathodic Protection of Pipelines in Handbook of Environmental Degradation of Materials*, ed. by M. Kutz (2012)
81. R.L. Liu, M.F. Hurley, A. Kvryan, G. Williams, J.R. Scully, N. Birbilis, Controlling the corrosion and cathodic activation of magnesium via microalloying additions of Ge. Sci. Rep. **6**, 28747 (2016)
82. N.K. Liyana, M.A. Fazal, A.S.M.A. Haseeb, Polarization and EIS studies to evaluate the effect of aluminum concentration on the corrosion behavior of SAC105 solder alloy. Mater. Sci.-Poland **35**(4), 694–701 (2017)
83. H. Wan, D. Song, C. Du, Z. Liu, X. Li, Effect of alternating current and Bacillus cereus on the stress corrosion behavior and mechanism of X80 steel in a Beijing soil solution. Bioelectrochemistry **127**, 49–58 (2019)
84. O.S.I. Fayomi, I.G. Akande, Corrosion mitigation of aluminium in 3.65% NaCl medium using hexamine. J. Bio-and Tribo-Corrosion **5**(1), 23 (2019)
85. J. Xu, Y. Bai, T. Wu, M. Yan, C. Yu, C. Sun, Effect of elastic stress and alternating current on corrosion of X80 pipeline steel in simulated soil solution. Eng. Fail. Anal. **100**, 192–205 (2019)
86. X. Wang, X. Song, Y. Chen, Z. Wang, L. Zhang, Study on corrosion and delamination behavior of X70 steel under the coupling action of AC-DC interference and Stress. Int. J. Electrochem. Sci. **14**(2), 1968–1985 (2019)
87. Y. Qing, Y. Bai, J. Xu, T. Wu, M. Yan, C. Sun, Effect of alternating current and sulfate-reducing bacteria on corrosion of X80 pipeline steel in soil-extract solution. Materials **12**(1), 144 (2019)
88. K. Tang, Stray alternating current (AC) induced corrosion of steel fibre reinforced concrete. Corros. Sci. (2019)
89. Y. Zhang, Q. Feng, L. Yu, C.M.L. Wu, S.P. Ng, X. Tang, Numerical modelling of buried pipelines under DC stray current corrosion. J. Electrochem. Sci. Eng. **9**(2), 125–134 (2019)
90. S.S.M. Tavares, J.M. Pardal, T.R.B. Martins, M.R. da Silva, Influence of sulfur content on the corrosion resistance of 17-4PH stainless steel. J. Mater. Eng. Perform. **26**(6), 2512–2519 (2017)
91. S. Lameche, R. Nedjar, H. Rebbah, A. Adjeb, Corrosion and passivation behaviour of three stainless steels in different chloride concentrations. Asian J. Chem. **20**(4), 2545 (2008)
92. H.J. Song, M.K. Kim, G.C. Jung, M.S. Vang, Y.J. Park, The effects of spark anodizing treatment of pure titanium metals and titanium alloys on corrosion characteristics. Surf. Coat. Technol. **201**(21), 8738–8745 (2007)
93. E.S.M. Sherif, A.A. Almajid, K.A. Khalil, H. Junaedi, F.H. Latief, Electrochemical studies on the corrosion behavior of API X65 pipeline steel in chloride solutions. Int. J. Electrochem. Sci. **8**, 9360–9370 (2013)

94. A. Fattah-alhosseini, A.R. Ansari, Y. Mazaheri, M. Karimi, Electrochemical behavior assessment of micro-and nano-grained commercial pure titanium in H_2SO_4 solutions. J. Mater. Eng. Perform. **26**(2), 611–620 (2017)
95. E.S. Ogawa, A.O. Matos, T. Beline, I.S. Marques, C. Sukotjo, M.T. Mathew et al., Surface-treated commercially pure titanium for biomedical applications: Electrochemical, structural, mechanical and chemical characterizations. Mater. Sci. Eng., C **65**, 251–261 (2016)
96. R. Yazdi, H.M. Ghasemi, C. Wang, A. Neville, Bio-corrosion behaviour of oxygen diffusion layer on Ti–6Al–4V during tribocorrosion. Corros. Sci. **128**, 23–32 (2017)
97. T. Yetim, An investigation of the corrosion properties of Ag-doped TiO_2-coated commercially pure titanium in different biological environments. Surf. Coat. Technol. **309**, 790–794 (2017)
98. A. Salehi, F. Barzegar, H.A. Mashhadi, S. Nokhasteh, M.S. Abravi, Influence of pore characteristics on electrochemical and biological behavior of Ti foams. J. Mater. Eng. Perform. **26**(8), 3756–3766 (2017)
99. Díaz, I., Pacha-Olivenza, M.Á., Tejero, R., Anitua, E., González-Martín, M.L., Escudero, M.L., García-Alonso, M.C., Corrosion behavior of surface modifications on titanium dental implant. In situ bacteria monitoring by electrochemical techniques. J. Biomed. Mater. Res. Part B: Appl. Biomater. **106**(3), 997–1009 (2018)
100. T. Bellezze, G. Roventi, R. Fratesi, Localised corrosion and cathodic protection of 17 4PH propeller shafts. Corros. Eng., Sci. Technol. **48**(5), 340–345 (2013)
101. W. Dou, J. Liu, W. Cai, D. Wang, R. Jia, S. Chen, T. Gu, Electrochemical investigation of increased carbon steel corrosion via extracellular electron transfer by a sulfate reducing bacterium under carbon source starvation. Corros. Sci. **150**, 258–267 (2019)
102. S. Li, L. Li, Q. Qu, Y. Kang, B. Zhu, D. Yu, R. Huang, Extracellular electron transfer of Bacillus cereus biofilm and its effect on the corrosion behaviour of 316L stainless steel. Colloids Surf., B **173**, 139–147 (2019)
103. T. Gu, R. Jia, T. Unsal, D. Xu, Toward a better understanding of microbiologically influenced corrosion caused by sulfate reducing bacteria. J. Mater. Sci. Technol. **35**(4), 631–636 (2019)
104. R. Jia, J.L. Tan, P. Jin, D.J. Blackwood, D. Xu, T. Gu, Effects of biogenic H_2S on the microbiologically influenced corrosion of C1018 carbon steel by sulfate reducing Desulfovibrio vulgaris biofilm. Corros. Sci. **130**, 1–11 (2018)
105. E.G. Zemtsova, A.Y. Arbenin, R.Z. Valiev, V.M. Smirnov, Modern techniques of surface geometry modification for the implants based on titanium and its alloys used for improvement of the biomedical characteristics. In *Titanium in Medical and Dental Applications* (2018), pp. 115–145
106. A.V.C. Sobral, W. Ristow Jr., S.C. Domenech, C.V. Franco, Characterization and corrosion behavior of injection molded 17-4 PH steel electrochemically coated with poly [trans-dichloro (4-vinylpyridine) ruthenium]. J. Solid State Electrochem. **4**(7), 417–423 (2000)
107. F.C. Moreira, R.A. Boaventura, E. Brillas, V.J. Vilar, Electrochemical advanced oxidation processes: a review on their application to synthetic and real wastewaters. Appl. Catal. B **202**, 217–261 (2017)
108. E. Haritha, S.M. Roopan, G. Madhavi, G. Elango, N.A. Al-Dhabi, M.V. Arasu, Green chemical approach towards the synthesis of SnO_2 NPs in argument with photocatalytic degradation of diazo dye and its kinetic studies. J. Photochem. Photobiol., B **162**, 441–447 (2016)
109. S. Ahmed, M.G. Rasul, R. Brown, M.A. Hashib, Influence of parameters on the heterogeneous photocatalytic degradation of pesticides and phenolic contaminants in wastewater: a short review. J. Environ. Manage. **92**(3), 311–330 (2011)
110. S.K. Tammina, B.K. Mandal, N.K. Kadiyala, Photocatalytic degradation of methylene blue dye by nonconventional synthesized SnO_2 nanoparticles. Environ. Nanotechnol. Monit. Manage. **10**, 339–350 (2018)
111. X. Deng, Q. Ma, Y. Cui, X. Cheng, Q. Cheng, Fabrication of TiO_2 nanorods/nanosheets photoelectrode on Ti mesh by hydrothermal method for degradation of methylene blue: influence of calcination temperature. Appl. Surf. Sci. **419**, 409–417 (2017)
112. M.M. Islam, S. Basu, Effect of morphology and pH on (photo) electrochemical degradation of methyl orange using TiO_2/Ti mesh photocathode under visible light. J. Environ. Chem. Eng. **3**(4), 2323–2330 (2015)

Chapter 4
Electrochemical Cells

Ikenna Chibuzor Emeji, Onoyivwe Monday Ama, Uyiosa Osagie Aigbe,
Khotso Khoele, Peter Ogbemudia Osifo, and Suprakas Sinha Ray

Abstract The main goal of this chapter is to present an overview of electrochemical cell operations. An electrochemical cell is devices that use a spontaneous chemical reaction to produce electricity or conversely use applied electricity to bring about non-spontaneous useful chemical reactions. The electroactive species in the ionic conductor (electrolyte) through mass transport reaches the electrode surface where Faradaic and Non-faradaic Processes occurs. A Faradaic process such as redox reaction at the electrode-solution interface gives rise to reduction or oxidation reaction. Fick's law gives the rate of diffusion of the oxidized or reduced species in terms of a concentration gradient. The electrified solution-electrode interface was modeled using Helmholtz compact layer model, Gouy-Chapman diffuse layer model, and the Stern model. Accepted definitions of certain physical quantities were also presented.

I. C. Emeji · O. M. Ama (✉) · P. O. Osifo
Department of Chemical Engineering, Vaal University of Technology, Private Mail Bag X021,
Vanderbijlpark 1900, South Africa
e-mail: onoyivwe4real@gmail.com

O. M. Ama · S. S. Ray
Department of Chemical Science, University of Johannesburg, Doornfontein 2028, South Africa

DST/CSIR National Centre for Nano-Structured Materials, Council for Scientific and Industrial
Research, Pretoria 0001, South Africa

U. O. Aigbe
Department of Physics, College of Science, Engineering and Technology, University of South
Africa, Pretoria 0001, South Africa

K. Khoele
Department of Chemical, Metallurgical and Materials Engineering, Tshwane University of
Technology, Pretoria, South Africa

© Springer Nature Switzerland AG 2020
O. M. Ama and S. S. Ray (eds.), *Nanostructured Metal-Oxide Electrode
Materials for Water Purification*, Engineering Materials,
https://doi.org/10.1007/978-3-030-43346-8_4

4.1 Introduction

A typical electrochemical cell contains two electrodes and an electrolyte [1]. They are devices which use spontaneous chemical reactions to produce clean electricity or, conversely, use electricity in form of electrical energy to bring about non-spontaneous useful chemical reactions. The electrochemical cell can be designed either to be divided or undivided. Undivided cells are more straightforward to set-up with both oxidation and reduction taking place within the same compartment. According to Watts et al. [2] the undivided cells setup (Fig. 4.1a) is designed in such a way that the products formed from redox reaction will not react with the electrodes. On the other hand, divided cells as shown in Fig. 4.1b requires a more specialized setup with electroactive species to be oxidized placed in the anodic compartment, while species to be reduced is placed in the cathodic compartment. The cathodic and anodic chambers are separated with a semi-porous membrane such as sintered glass frit, porous porcelain, polytetrafluoroethene, an ion-selective membrane or salt bridge. The purpose of the divided cell is to "separate" the chemistries at the two electrodes [3] hence permitting the diffusion of ions via the membrane and restricting the flow of the products and reactants.

Other functions of the porous membrane are to complete the electric circuit and maintain electrical neutrality on both sides of the electrode compartment. The two resulting half-cells may utilize the same or different types of electrolytes. Electrolyte, therefore, is an ionic substance that permits ions to migrate between the electrode compartments, thereby sustaining the system's electrical neutrality. The two types of electrolytes are strong and weak electrolyte. Those electrolytes which completely ionize or dissociate into ions are known as strong electrolytes. Some of the examples of strong electrolytes are hydrochloric acid (HCl), sodium hydroxide (NaOH), and potassium sulfate (K_2SO_4). Weak electrolytes, on the other hand, are those electrolytes that dissociate partially in an aqueous medium. Some of the examples of

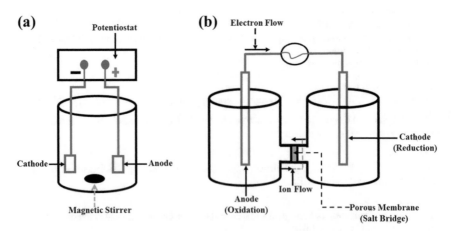

Fig. 4.1 a Undivided cell. b Divided cell

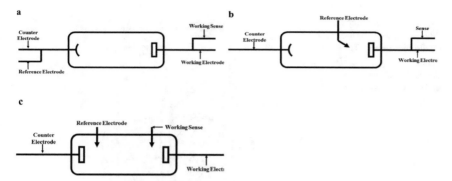

Fig. 4.2 a Overview of 2-electrode setup. **b** Overview of 3-electrode setup. **c** Overview of 4-electrode setup

weak electrolytes are Acetic acid (CH_3COOH), carboxylic acid (H_2CO_3), ammonium hydrosulphide (NH_4OHH_2S), etc. Electrochemical reactions occurring in a cell are between the electrolyte, electrodes and an external substance (as in fuel cells that may use hydrogen gas as a reactant). Electrodes are used to perform specialized roles in electrolytic cells; hence cells can be configured or set up to have two, three or four electrodes, depending on the type of experiment to be conducted. The most basic form of cell configuration is the two-electrode cell configuration, which usually has the electrode under investigation as the working electrode (WE) and the electrode necessary to close the electrical circuit, the counter electrode (CE). The electrode where the chemistry of interest occurs is the WE. The physical set-up for two electrode mode is shown in Fig. 4.1a with working (W) and working sense (WS) leads connected to the working electrode (WE) called the cathode and Reference (R) and counter (C) leads connected to the second electrode called the anode.

In this type of setup, the counter electrode has two major responsibilities. It completes the circuit permitting charges to be circulated throughout the cell and it sustains the same interfacial potential despite the flow of free electrons. The role of circulating current with constant voltage supply are therefore, better utilized by two separate electrodes. Thus, this kind of setup is usually employed when one wants to examine and analyze properties such as electrolyte conductivity or to characterize semiconductor devices. For three-electrode cell configurations, the cell setup consists of the working electrode (WE), a counter electrode (CE), and a reference electrode (RE) as depicted in Fig. 4.3. In this setup, the RE is most often positioned in such a way as to measure and control the working electrode potential (which has both Working and Working Sense leads attached (Fig. 4.2b), without allowing the passage of free flow electrons. The RE are expected to sustain the electrochemical potential value at low current density. Furthermore, since the RE allows the passage of so small current, the potential drop across the space separating the reference and the working electrode (iRU) is often very negligible. Hence, three-electrode system gives much more stable reference potential value than others with an additional compensation of iR drop in solution. Therefore, in the three-electrode setup, the major role of the CE

Fig. 4.3 Simple schematic diagram of 3-electrode cell configuration with E_A as the applied voltage, WE, CE and RE as the working, counter and reference electrodes respectively

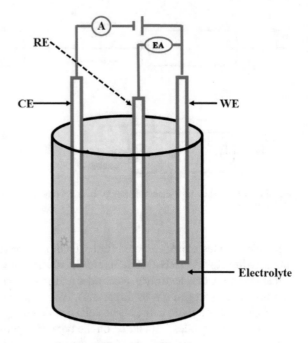

is to allow the passage of all free electrons required to stabilize any current detected at the working electrode.

Hence, maximum potentials at the CE are needed so as to achieve such a task. The most uncommon cell configuration is the four-electrode cell configuration, which has the Working Sense lead decoupled from the working electrode, as shown in Fig. 4.2c. In a 4-electrode mode, working electrode potentials are not being measured but rather the effect of an applied current or some barrier on the solution itself were measured. A 4-electrode cell setup is generally used to analyze processes occurring within the electrolyte, between two measuring electrodes separated by a membrane [4].

1. **Types of Electrochemical Cells**

The two main types of electrochemical cells and they are galvanic cells (also called voltaic cells) and electrolytic cells.

4.1.1 Galvanic Cells

A galvanic cell is a device that utilizes spontaneous oxidation-reduction reaction to produce an electric current; hence it is also called voltaic cell [5]. Spontaneous redox reactions are exothermic reactions. They change the energy liberated by a spontaneous reduction and oxidation reaction into electrical energy used to do work. Hence, in galvanic cells, the redox half-reactions typically occur in different compartments called half-cells. As depicted in Fig. 4.1b, the porous barrier connects the two half-cell allowing ions to migrate between the two chambers while also retaining their electrical neutrality. The potential difference between the two electrodes (voltage) enables electrons to migrate from the reduced to the oxidized through an outer circuit, hence producing an electric current.

The metal that is more readily oxidized serves as an anode with oxidation half-reaction occurring there while the cathode is the electrode at which the reduction half-reaction occurs. Using zinc (Zn) and copper (Cu) for illustration, the basic principles of galvanic cell are represented in the following cell notation [6]

$$Zn|ZnSO4\ (aq)||CuSO4\ (aq)|Cu \tag{4.1}$$

(Cell notation has its cathode displayed on the right-hand side and the anode on the left)

$$Zn(s) + Cu^{2+}(aq) \rightarrow Zn^{2+}(aq) + Cu(s) \tag{4.2}$$

The ionic flow (electricity resulting from positive charge) is migrating from left to right once zinc rod is immersed into its aqueous electrolytic solution of zinc (II) sulfate. As the reaction continues, the zinc rod melts with the appearance of massive copper metal. These conversions occur automatically, with all the energy released in the form of heat that can be dissipated. Conclusively, in Galvanic Cell (aka Voltaic Cell), a chemical redox reaction is spontaneous as the change in Gibb's energy is less than zero (ΔG is <0), hence electrode potential is greater than zero ($E^0 > 0$). Electrons, therefore, migrate from the anode, the negative terminal to the cathode, the positive terminal (where electrons are gained). Examples of Galvanic cells are (a) Dry cell (i.e. Non-rechargeable batteries) and (b) Daniel cell.

4.1.2 Electrolytic Cell

The electrolytic cell, on the other hand, is an electrochemical cell that drives a non-spontaneous redox reaction through the application of an external source of electrical energy. In electrolytic cells, only a single compartment is employed in its applications. This kind of cell is mostly used to disintegrate chemical compounds, in a series of chemical operations called electrolysis. Examples of where this kind of cell is used are in rechargeable batteries, to divide solid metals from its metallic compound, to divide other chemical compounds such as water, to electroplate metals and too impressed current cathodic protection. The main differences between galvanic and electrolytic cells are outlined in Table 4.1. However, all electrolytic cells contain three main parts: two solid electrode conductors in contact with its ionic conductors (electrolytes). The negative charge electrode is always the cathode while the positively charged is the anode. These two electrodes are manufactured from substances that take part in the chemical reaction of which zinc, copper, and silver are examples. Also, like graphite,

Table 4.1 Correlation between galvanic and electrolytic cells

Galvanic cell	Electrolytic cell
ΔG is <0, E^0 cell >0	$\Delta G > 0$, E^0 cell <0
A Galvanic cell transforms chemical energy into electrical energy	An electrolytic cell transforms electrical energy into chemical energy
A spontaneous redox reaction occurs here which gives rise to the generation of electrical energy	Electrical energy from external source initiates redox reaction, hence reaction is not spontaneous
The cell is configured to have the two electrodes in different half-cells compartments, linked together via salt bridges or porous partition	The cell is configured to have both electrodes in the same compartment having the same electrolyte
Anode is the negative pole while cathode is the positive pole. An oxidation reaction occurs at the anode, while that at the cathode is a reduction	Cathode is the negative pole while anode is the positive pole. An oxidation reaction occurs at the anode, while that at the cathode is a reduction
Oxidized species are the source of electrons supplied and they move from anode to the cathode	External source supplies the electrons and forced them onto the cathode and eventually come out through the anode

silicon, or platinum, they can also be produced from chemically inert materials, thus they are generally regarded as active electrodes. First, the battery from external source supply's electrical energy which sends the electrons onto the cathode, making it negatively charged.

Pulling out electrons from the anode makes it positively charged. Once this occurs, a Redox reaction is initiated. Hence, an oxidation reaction occurs at the anode, allowing electrons to migrate freely, and becoming attracted toward the positive electrode. Consequently, electrons accumulate at the cathode where reduction takes place.

4.1.3   Electrochemical Cell Electrodes

Electrodes are typically good electric conductors, carrying currents to non-metallic solids, liquids or gases. In solid conductors, current is carried by electrons. Some examples of solid conductors are zinc, carbon, and iron. In liquids and gases, current is being carried by molecules which have acquired either positive or negative charges called ions. Examples include hydrogen ion (H^+), hydroxyl ion (OH^-) etc. According to Inzelt [7], there are numerous ways of classifying electrodes, hence enumerated. The electrode where reduction chemical reaction takes place is called the cathode and that where oxidation chemical reaction occurs is called the anode. Classifying electrode based on whether migratory ionic species crosses the boundary at the interface or not, we have non-polarizable electrode which represents electrodes that allows mass transfer at the interface and polarizable electrode which does not allow any charge transfer at their interface. Based on the electrode size, electrodes can consequently be classified into microelectrodes, microelectrodes, and ultra-microelectrodes. In terms of functional grouping, electrodes are classified as working, counter and reference electrodes.

4.1.4 Working Electrodes (WE)

This is the electrode where the reaction of interest takes place. Its usage changes from one experiment to another, to provide surface for the adsorption of electroactive species of interest [8]. The selection of WE depend either on the reduction and oxidation performance of the targeted species and the background current produced over the potential area. Other factors which can be examined when chosen WE, includes low-cost, electrical conductivity, stability, availability, surface modification and toxicity. There are different varieties of materials which can be used as WE for electroanalysis includes mercury, carbonaceous materials, noble metals, etc.

4.1.5 Counter or Auxiliary Electrode (CE)

Auxiliary electrode is another name for counter electrode. It is therefore, the electrode used to complete the electrical circuit in the electrochemical cell. Its major duty is to provide a different path for the current to pass so that only an infinitesimal amount of current migrates through the reference electrode. Commonly, they are called inert materials as they do not take part in an electrochemical reaction. Free flow electrons which is termed current, moves between the WE and the CE. To make certain that the dynamics of chemical reaction taken place at the CE do not obstruct those reaction happening at the WE, the surface area of the CE must be larger than the surface area of the WE. When observing a reduction reaction at the WE, a simultaneous oxidation reaction takes place at the CE. Therefore, chemically inactive CE were chosen for the electrode to be as inert as possible.

4.1.6 Reference Electrode

RE is an electrode with constant potential and negligible current. Their potential is normally used to make reference with the potential of other electrodes. Laboratory most commonly used reference electrodes, in aqueous media are silver-silver chloride Ag/AgCl electrode, the saturated calomel electrode (SCE) and the standard hydrogen electrode (SHE). These RE's, in solution are mostly isolated by a porous membrane. RE's derived from Ag^+/Ag couple is the most often used in non-aqueous solvents, hence, a very high input impedance on the electrometer (>100 GOhm) together with the CE, enables current flowing through the reference electrode to be maintained in close proximity to zero (ideally, zero).

The half-reaction for the AgCl/Ag reference electrode is as follows:

$$AgCl\,(s) + e^- \rightarrow Ag(s) + Cl^-(aq) \tag{4.3}$$

The electrode half-cell notation gives:

$$Ag(s)/AgCl(s)//KCl\,(aq), AgNO3\,(aq) \tag{4.4}$$

The half reaction that occurs inside the saturated calomel (SCE) reference electrode is as follows

$$Hg_2Cl_2(s) + 2e^- \rightarrow 2\,Hg(I) + 2\,Cl^-\,(aq)$$

The shorthand notation for the SCE half-cell is as follows:

$$Pt(s)/Hg(I)//Hg_2Cl_2(s)/KCl\,(aq).$$

4.2 Electrode Potential (E^0)

The electrode potential, E (SI unit is Volts) is the electric potential difference of an electrochemical cell. It is the potential value delivered by an electrode to its interfacial boundary with the electrolyte. In an electrochemical cell, each electrode exacts an electric potential at the interface with the electrolyte due to charged species which are transported over the interface. The difference between the two electrode potentials gives the cell potential:

$$E^0_{cell} = E^0_{cathode} - E^0_{anode} \tag{4.5}$$

When current is moves across any given electrochemical cell, the resulting reference electrodes potential should be steady for the purpose of maintaining a precise potential value for the electrode under analysis. In accordance with the recommendations of IUPAC, Standard hydrogen electrode (SHE) should be used as the reference point electrode. Hence, hydrogen is assigned a reference standard electrode potential of zero which all other electrodes are measured against. When SHE acts as the cathode, the following half-cell reaction occurs

$$2H^+(aq) + 2e^- \rightarrow H_2(g)\, E^0_{cathode} = 0.000\,V \tag{4.6}$$

If the SHE is connected to an electrode in another half-cell, the cell potential (the difference in potential energy) can be used to determine the standard electrode potential of the other half cell. Therefore, the cell potential of an electrochemical cell is the difference in voltage between the cathode (final stage) and the anode (the initial stage) as given in (4.5). E_{cell}^0 Is positive for reactions occurring spontaneous and negative for a non-spontaneous reaction?

For example, calculating the electrode potential of zinc, one need to connect zinc half-cell to the hydrogen electrode. Thus, zinc being higher in the electrochemical series (Fig. 4.4) will have excess electrons (i.e. negative ions) which need to be discharged. Zinc electrode becomes more electro-negative (anode) so as to discharge the electrons. Hydrogen becomes more positive (cathode) so as to gain an electron as they are released.

$$Zn^{2+} + 2e^- \rightarrow Zn \text{ (s) (anodic half-reaction)} \tag{4.7}$$

$$2H^+(aq) + 2e^- \rightarrow H_2(g) \text{ (cathodic half-reaction)} \tag{4.8}$$

Using the formula,

$$E_{cell}^0 = E_{cathode}^0 - E_{anode}^0 = E_{H_2}^0 - E_{Zn}^0$$
$$E_{cell}^0 = 0 - 0.76 = -0.76$$

Like zinc, copper possesses an excess of electrons (i.e. negative ions) and are suppose become more electronegative but because of its position in the electrochemical series, being below hydrogen the reference electrode, it will serve as the cathode with hydrogen serving as the anode.

Illustration question Using similar method as in above, calculate the standard potential of the given reaction as it occurs in an electrochemical cell at 25 °C (equation is balanced).

$$3Pb^{2+} \text{ (aq)} + 2Cr \text{ (s)} \rightarrow 3Pb \text{ (s)} + 2Cr^{3+} \text{ (aq)} \tag{4.11}$$

Reduction half reactions	E^0(V)
Pb^{2+}(aq) + 2e$^-$ $\Rightarrow$ Pb (s)	−0.13
Cr^{3+}(aq) + 3e$^-$ $\Rightarrow$ Cr (s)	−0.73

Solution Comparing Pb^{2+} and Cr^{3+} to the position of standard referencing hydrogen in Fig. 4.4, Cr^{3+} is farther away, so possesses an excess of electrons (i.e. negative

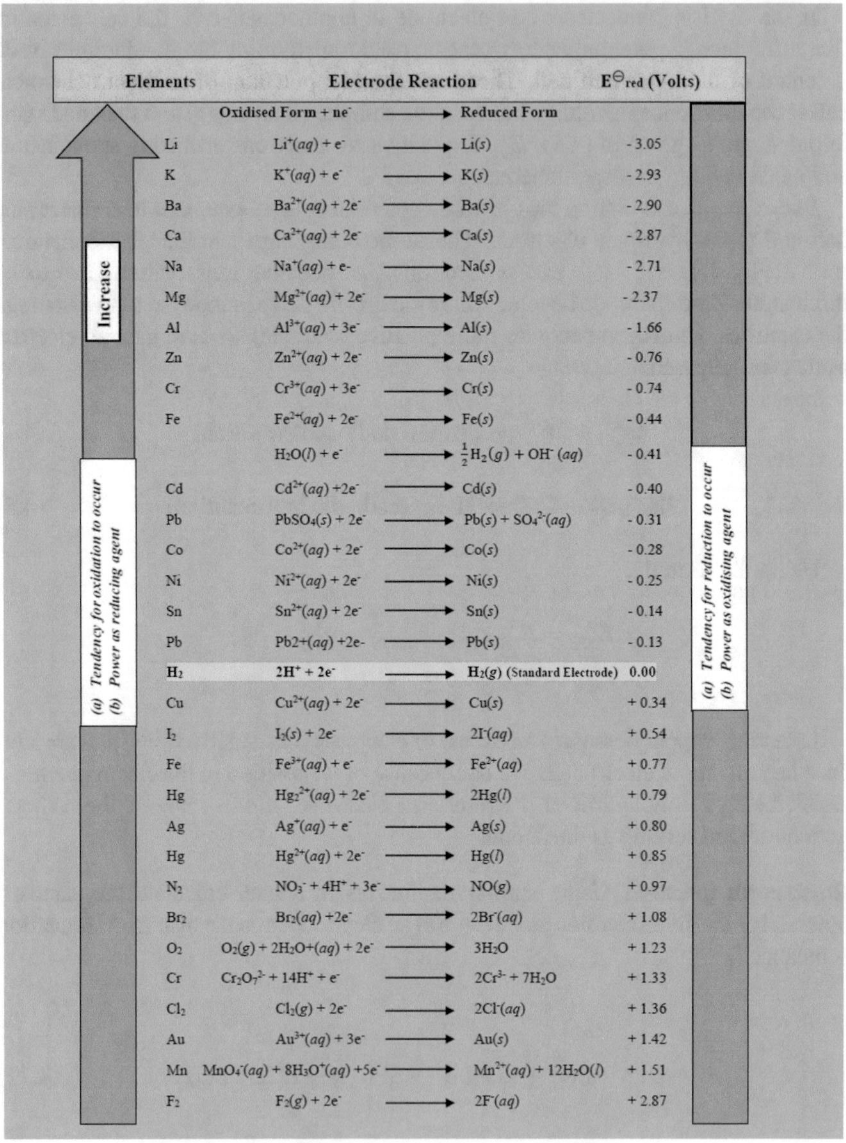

Elements	Electrode Reaction	$E^{\ominus}_{red}$ (Volts)
	Oxidised Form + ne⁻ $\longrightarrow$ Reduced Form	
Li	$Li^+(aq) + e^- \longrightarrow Li(s)$	- 3.05
K	$K^+(aq) + e^- \longrightarrow K(s)$	- 2.93
Ba	$Ba^{2+}(aq) + 2e^- \longrightarrow Ba(s)$	- 2.90
Ca	$Ca^{2+}(aq) + 2e^- \longrightarrow Ca(s)$	- 2.87
Na	$Na^+(aq) + e^- \longrightarrow Na(s)$	- 2.71
Mg	$Mg^{2+}(aq) + 2e^- \longrightarrow Mg(s)$	- 2.37
Al	$Al^{3+}(aq) + 3e^- \longrightarrow Al(s)$	- 1.66
Zn	$Zn^{2+}(aq) + 2e^- \longrightarrow Zn(s)$	- 0.76
Cr	$Cr^{3+}(aq) + 3e^- \longrightarrow Cr(s)$	- 0.74
Fe	$Fe^{2+}(aq) + 2e^- \longrightarrow Fe(s)$	- 0.44
	$H_2O(l) + e^- \longrightarrow \frac{1}{2}H_2(g) + OH^-(aq)$	- 0.41
Cd	$Cd^{2+}(aq) + 2e^- \longrightarrow Cd(s)$	- 0.40
Pb	$PbSO_4(s) + 2e^- \longrightarrow Pb(s) + SO_4^{2-}(aq)$	- 0.31
Co	$Co^{2+}(aq) + 2e^- \longrightarrow Co(s)$	- 0.28
Ni	$Ni^{2+}(aq) + 2e^- \longrightarrow Ni(s)$	- 0.25
Sn	$Sn^{2+}(aq) + 2e^- \longrightarrow Sn(s)$	- 0.14
Pb	$Pb^{2+}(aq) + 2e^- \longrightarrow Pb(s)$	- 0.13
H₂	$2H^+ + 2e^- \longrightarrow H_2(g)$ (Standard Electrode)	0.00
Cu	$Cu^{2+}(aq) + 2e^- \longrightarrow Cu(s)$	+ 0.34
I₂	$I_2(s) + 2e^- \longrightarrow 2I^-(aq)$	+ 0.54
Fe	$Fe^{3+}(aq) + e^- \longrightarrow Fe^{2+}(aq)$	+ 0.77
Hg	$Hg_2^{2+}(aq) + 2e^- \longrightarrow 2Hg(l)$	+ 0.79
Ag	$Ag^+(aq) + e^- \longrightarrow Ag(s)$	+ 0.80
Hg	$Hg^{2+}(aq) + 2e^- \longrightarrow Hg(l)$	+ 0.85
N₂	$NO_3^- + 4H^+ + 3e^- \longrightarrow NO(g)$	+ 0.97
Br₂	$Br_2(aq) + 2e^- \longrightarrow 2Br^-(aq)$	+ 1.08
O₂	$O_2(g) + 2H_3O^+(aq) + 2e^- \longrightarrow 3H_2O$	+ 1.23
Cr	$Cr_2O_7^{2-} + 14H^+ + e^- \longrightarrow 2Cr^{3+} + 7H_2O$	+ 1.33
Cl₂	$Cl_2(g) + 2e^- \longrightarrow 2Cl^-(aq)$	+ 1.36
Au	$Au^{3+}(aq) + 3e^- \longrightarrow Au(s)$	+ 1.42
Mn	$MnO_4^-(aq) + 8H_3O^+(aq) + 5e^- \longrightarrow Mn^{2+}(aq) + 12H_2O(l)$	+ 1.51
F₂	$F_2(g) + 2e^- \longrightarrow 2F^-(aq)$	+ 2.87

Increase

(a) Tendency for oxidation to occur (b) Power as reducing agent

(a) Tendency for reduction to occur (b) Power as oxidising agent

Fig. 4.4 Standard potential of some common metals and non-metals

ions) which need to be released. Therefore, In order to release those electrons, Cr^{3+} become more electro-negative (cathode). Electrons, therefore, tend to move away from the negative electrode, hence are discharged. They are accepted by the positive electrode Pb^{2+} (anode) which is more electro-positive.

Applying (4.5), we have $E^0_{cell} = E^0_{cathode} - E^0_{anode}$

$$E_{cell}^0 = E_{Pb}^0 - E_{Cr}^0 = -0.13 - (-0.73) = +0.60\text{ V}$$

Since E^0 of the cell is positive, cell reaction is spontaneous, and the type of cell is Galvanic Cell.

4.3 Nernst Equation

In most electrochemical cells, the temperature of the cell varies. The concentration of the reactants and that of the products are also varied as they are not normally equal to 1 M (or 1 atm, for gases). Hence, the type of reduction reaction which takes place at the surface of the electrode in an electrochemical series is of the form:

$$O + ne^- \rightleftharpoons R \qquad (4.12)$$

where O represents the oxidized species, R represents the reduced species and n gives the number of electrons swapped between O and R. The equation linking the oxidized and reduced species in terms of concentration, with free energy (J mol^{-1}) is given by

$$\Delta G = \Delta G^0 + RT \ln \frac{[R]}{[O]} \qquad (4.13)$$

Or

$$\Delta G = \Delta G^0 + RT \ln(Q) \qquad (4.14)$$

where Q is the reaction quotient or activity. However, the authentic cell potential is different from the standard cell potential, the maximum work that can be done in a cell, is equal to change in Gibbs free energy. Hence,

$$\Delta G = -nFE \qquad (4.15)$$

where n represents the number of moles of electrons transported in the reaction (per mole of reactant or product), F represents the Faraday's constant $= 96,500$ C/Mol. and E gives the peak potential between the two electrodes, which can also be referred to as open circuit potential (OCP) or equilibrium potential. If both the reacting species and the product species have unit activity, i.e. at standard conditions, then the above equation became

$$\Delta G^0 = -nFE^0 \qquad (4.16)$$

In this situation, the potential is called the standard electrode potential or standard potential, and it is related to the standard free energy change (ΔG^0) by the above equation. Therefore, the expression relating the potential and concentration of electroactive species taken part in any cell reaction at equilibrium is known as the Nernst equation. By substituting (4.15), and (4.16) into (4.13), we get the Nernst equation, hence

$$-nFE = -nFE^0 + RT \ln\left(\frac{[R]}{[O]}\right)$$

$$E = E^0 - \frac{RT}{nF} \ln \frac{[R]}{[O]}$$

$$E = E^0 + \frac{RT}{nF} \ln \frac{[O]}{[R]}$$

$$E = E^0 + \frac{0.059}{n} \ln \frac{[O]}{[R]} \text{ at } 25\,^{\circ}\text{C} \tag{4.17}$$

Under standard conditions, all concentrations are equivalent to 1 m or 1 atm from the Nernst equation. The expression also indicates that the electrode potential of a cell is dependent on the reaction quotient Q of the reaction. As the reduction and oxidation reaction proceeds, reacting species are used up, hence, decreasing the concentration of the reactants. Conversely, the concentration of the products formed increases. As this happens, the cell potential slowly decreases until the reaction reaches equilibrium. At equilibrium, $\Delta G = 0$, also $E = 0$. The reaction quotient Q = equilibrium constant, K. then (4.14)

$$\Delta G = \Delta G^0 + RT \ln(Q) \equiv \Delta G^0 + RT \ln \frac{[R]}{[O]} \text{ become}$$

$$O = \Delta G^0 + RT \ln K$$

$$\Delta G^0 = -RT \ln K \tag{4.18}$$

Nernst equation then gives the standard electrode potential of the cell as,

$$-nFE^0 = -RT \ln K,$$

$$\Rightarrow E^0 = \frac{RT \ln(K)}{nF}$$

But, when Q equals one, its logarithm gives zero, then the cell potential equals the standard cell potential. Hence:

i.e. When $Q = 1$, then $\frac{[R]}{[O]} = 1$. Its logarithm gives. $\ln Q = \ln \frac{[R]}{[O]} = \ln 1 = 0$.

Hence, $E = E^0$

4.4   Faradaic and Non-faradaic Processes

Two types of surface operations occur at the electrode surface. One is redox reaction which involves the migration of electrons at the electrode-solution interface. Since such reactions are governed by Faraday's law (which states that the number of moles of electron, produced or consumed during an electrode process, is proportional to the quantity of electricity passing through the electrode), they are called faradaic processes. By Faraday's law, we have:

$$Q = nFN, \tag{4.19}$$

where (Q) represents the total charge or quantity of electricity in the experiment, N is the number of molecules electrolyzed and n is the number of electrons transferred. Therefore, Faradaic current is given by

$$\frac{dQ}{dt} = i = nF\frac{dN}{dt} \tag{4.20}$$

Under certain given conditions, the electrode-solution interface can exhibit a set of potentials which does not support charge-transfer reactions because such reactions are disadvantaged. However, unit operations such as adsorption, desorption, movement of electrolytic ions and reorientation of solvent dipoles can occur, changing the arrangement of the electrified electrode-solution interface. Although during such processes, no charge crosses the electrode-electrolyte interface, but external current can flow with irregular potential, electrode area or solution composition. Such processes are called non-Faradaic processes. Hence, both Faradaic and non-Faradaic operations occur when electrode reactions take place. Generally, we understood that in an electrode reaction, the exchange of electron at the interface between electrode and solution species result in a Faradaic current flow. Such current arising is controlled by:

(1)   The rate at which electron migrates between the metal electrode and the bulk electroactive species and
(2)   The mass transfer processes occurring at the interface by the bulk electroactive species.

Therefore, the transport of single electron between two species of (O) and (R) gives the reductive and oxidative current flowing as:

$$i_c = -FAK_{RED}[O]_o \tag{4.21}$$

$$i_a = -FAK_{OX}[R]_o \tag{4.22}$$

where i_c and i_a are the reduction and oxidation current, K_{RED} and K_{OX} are the rate constants of the electron transfer, $[O]_o$ is the surface concentration of the reactants. Therefore, the corresponding reaction rates are predicted by the equation of the form

$$K = Ze^{\frac{-\Delta G}{RT}} \tag{4.23}$$

$$K_{RED} = Ze^{\frac{-\Delta G_{RED}}{RT}} \tag{4.24}$$

$$K_{OX} = Ze^{\frac{-\Delta G_{OX}}{RT}} \tag{4.25}$$

So, for a single applied voltage, the free energy outline seems to be exactly the same in quality as the corresponding chemical processes. On the assumption that the effect of voltage application gives a linear relationship on free energy, then the rate constants can be given as.

$$K_{RED} = Ze^{\frac{-\Delta G_{red\,no\,voltage}}{RT}\frac{-\alpha FV}{RT}} \tag{4.26}$$

$$K_{RED} = Ze^{\frac{-\Delta G_{ox\,no\,voltage}}{RT}\frac{(1-\alpha)FV}{RT}} \tag{4.27}$$

Therefore, the obtained reveals that for electron transfer processes, the rate constant is proportional to the exponential value of the applied voltage. Hence, the rate of electrolysis can become different by changing the applied voltage.

4.4.1 Lectrochemical Resistance

Electrochemical cells are constructed of various materials, such as the wire, the electrolytic solutions, the electrodes, and the containers. All of these materials together cause the cell to experience some sort of resistance. However, charges are transported in the electrolyte through the motion of ions (positive and negative). As the ions migrate from one point to another in the solution, ionic resistance occurs. The resistance of an ionic solution is therefore determined by the geometry of the area in which current is carried, the ionic concentration and type of ions, temperature. Charge transport can also occur in the electrode through the motion of electrons. Electronic resistance occurs along the electrode and external circuit as an electron is transferred. This type of resistance can be minimized by using electron conducting materials. However, over a given cross-sectional area A and length l carrying a uniform current, the resistance is expressed as:

$$R = \rho\frac{l}{A} \tag{4.28}$$

where ρ represents the solution resistivity. Solution conductivity, κ, is regularly used in resistivity calculations, hence the expression:

$$R = \frac{1}{K}\frac{l}{A} \quad \Rightarrow \quad K = \frac{l}{RA} \tag{4.29}$$

Standard chemical handbooks list conductivity (κ) values of specific solutions as a function of concentration [9]. The units for conductivity (κ) are Siemens per meter (S/m). Multiplying Siemens with ohms gives unity by definition, so $1\ S = 1/\Omega$. Generally, theses resistance causes the cell to generally lose some of its potentials. What remains of the cell's voltage after subtracting the voltage loss due to internal resistance is called the terminal potential difference. This is expressed as:

$$V_{t,p,d} = emf - Ir \tag{4.30}$$

In this equation, $V_{t.p.d.}$ is the terminal potential difference, emf is the electromotive force, or ideal amount of energy (in volts) provided by the cell before there is resistance, I, represents current in amperes, and represents the internal resistance in ohms.

4.4.2 Electrode Reaction

All electrochemical reactions can either be referred to as anodic or cathodic reactions. Electrochemical reactions happening at the interface between the electrode and its electrolyte are frequently examined thoroughly, in electrochemistry, with the application of electrode potential. These chemical reactions actually resulted from the heterogeneous migration of electrons between the electrode and the electroactive species in solution [10]. Hence, a standard electrode reaction requires an electron to be transferred between an electrode and the bulk electroactive species. It can also be referred to as electrolysis and produces positive or negative currents depending on the type of half-reaction occurring at the surface of the electrode. Electrode reaction may occur as a single electron-transfer step, or as a sequence of two or more steps. The half-reactions occurring at the surfaces of the electrodes are typically expressed as follows:

$$O + me - \underset{OXIDATION}{\overset{REDUCTION}{\longleftrightarrow}} R \tag{4.31}$$

where: O (Reactant) gains an electron and became reduced (GER), R (Product) losses electron and became oxidized (LEO), Oxidation is the loss of electron/increase in oxidation number and Reduction is the gain of electron/decrease in oxidation number.

In all these electrochemical reactions, chemical bonds are broken, and new bonds are formed as with all types of chemical reactions. However, the electrode where reduction reactions occur is called the cathode and the electrode where oxidation reactions occur is called anode. Whether an oxidation or reduction reaction can take place at the interface is determined by the relative energy of the electrons within the working electrode compared with the energy of the ionic species in the electrolyte. By controlling the applied potential on the working electrode, the surface reaction of the electrode can be controlled. The mechanism or pathway of electrode reactions generally involves many steps as shown in Fig. 4.5. The steps include, transportation of reactant electroactive species (O) to the electrode surface, this is termed mass transport, adsorption of the electroactive species on the surface of the electrode, electrode reaction which involves exchange of electron through quantum mechanical tunnelling between the electrode and the electroactive species closer to the electrode (tunnelling distances according to Sun et al. [11] is <2 nm), products-R desorption from the electrode surface, transport of R-product away from the electrode surface to allow fresh reactant to accumulate on the electrode.

However, at low over-potentials, the electron transfer step is often the rate-limiting step, while at higher over-potentials mass transport becomes the limiting factor. Generally, the fundamental event in electrochemical reaction occurring on the surface of electrode in the electrochemical cell is the electron exchange between the electrode and a chemical species in solution occurring at their interface. This exchange enables

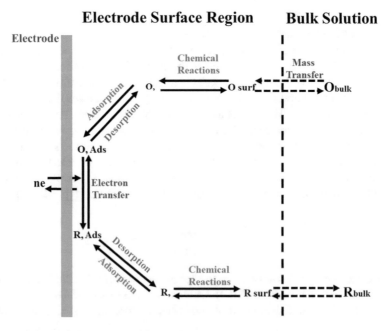

Fig. 4.5 Pathway of a general electrode reaction

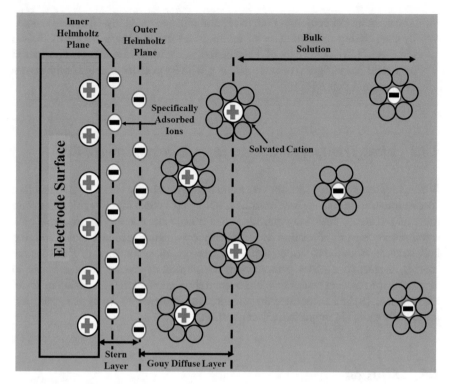

Fig. 4.6 Electric double layer (EDL) showing Helmholtz compact layer, Gouy-Chapman diffuse layer and Stern model

the interface to be electrified (i.e. current flows around) forming a double layer, called electric double layer (EDL). Therefore, Helmholtz compact layer model, Gouy-Chapman diffuse layer model and Stern model as depicted in the diagram in Fig. 4.6, were used to explain EDL [12], hence:

(a) Helmholtz model of the EDL as shown in Fig. 4.6, comprises of all species that are specifically adsorbed on the electrode surface. Hence, on the electrode surface, the excess adsorbed ions line up in a plane called the inner Helmholtz plane (IHP). Ions in the IHP have lost their solvation shell partially or completely and hence are in direct contact with the electrode.

(b) an outer Helmholtz layer (OHP), which comprises of hydrated (solvated) cations. These ions are closest to the electrode surface but are not specifically adsorbed to the electrode surface. Hence, ions in the OHP are hydrated with their solvation shell intact.

(c) Gouy-Chapman model assumes an outer "diffusion layer" containing excess cations or anions extending from the OHP to the bulk solution through a certain thickness, which is determined by the ion concentration.

(d) The stern model combines both the Helmholtz and the Gouy-Chapman models
      by considering the finite size of the ions and introducing a "plane of closest
      approach" [13]. In the bulk of the ionic liquid, cations and anions are considered
      as a solid round figure, of same radius r, with its polarity situated at the centre
      of the round figure [12].

4.4.3   Mass Transport Processes in an Electrochemical Cell

Whenever there is charge transfer process at the surface of the electrode, the reacting
species are used-up as they are consumed, setting up a concentration gradient within
the solution. Under diffusion-controlled conditions, more reacting species from bulk
solution moves in the direction of the electrode as formed product moves into the
bulk solution from the electrode surface. Therefore, mass transport of electroactive
species in electrochemistry demonstrates the motion of species from one area to
another in the solution due to species concentration gradient in the solution mixture.
Elgrishi et al. [8] hence, reported that the three fundamental modes of mass transport
are (1) diffusion, (2) migration, (3) convection.

4.5   Diffusion

Diffusion is the transport of particles as a result of the local difference in the chemical
potential [14]. On another word, diffusion is simple the random motion of a species
from an area of higher concentration to another area of lower concentration and it
must occur whenever bonds are broken or formed at the electrode surface. Diffusion-
controlled electrode reaction arises when the molecular species close to the surface
of the electrode are used-up, thereby reducing its amount near the electrode surface,
hence more species diffuses in from the bulk solution to replenish it. Similarly, when
a species is continually generated at an electrode, it begins to diffuse away from the
electrode, where its concentration is zero. The rate of diffusion depends upon, the
solution concentration gradient, and on the diffusion coefficient, D of electroactive
species fixed temperature. Therefore, the movement of intermolecular species as
effected by concentration gradient is described by Fick's first law which states that
the flux $J_A(x, t)$ of substance A, moving from higher concentration region to lower
concentration region, has a magnitude which is proportional to the concentration
gradient $\partial C_A(x, t)/dx$

$$J_A(x, t) = -D_A \frac{\partial C_A(x, t)}{\partial x} \qquad (4.32)$$

If the diffusion coefficient is self-sufficient and does not depend on position, then Fick's Second Law may clarify the situation, hence:

$$\frac{\partial C_o}{\partial t} = D_o \left\{ \frac{\partial^2 C_o}{\partial x^2} \right\} \tag{4.33}$$

Diffusion of reactants and products in electrolytes should be considered in many types of interfacial processes. Diffusion of atoms on the electrodes may control the rate on the deposition and the value of the diffusion coefficient would, of course, vary from system to system.

4.5.1 Migration

The movement of charged particles in response to a local electric field is called migration [15]. This electrostatic phenomenon emerged due to the application of voltage on the electrodes. Therefore, electroactive species near this electrified interface will moreover be attracted or repelled from the electrode surface by electrostatic forces. However, ion solvation effects make intermolecular movement within real solution very difficult to determine correctly. As a result, it has become unavoidably essential to add inert electrolyte in large quantity to the cell solution so as to shield the molecular species of any interest from migratory effect. The inert electrolyte commonly called the supporting electrolyte offsets the effect of intermolecular forces linking the working electrode and the molecular species by subduing the transported quantity of the reactants. This electrolyte can also act as a conductive solution to assist in the movement of current.

4.6 Convention

Convection is the movement of solution species due to externally controlled forces. Forced movement of bulk electroactive species by rotation or vibration or automated stirring give rise to forced convention. On the other hand, natural convection arises in solution due to changes in temperature and density. Fluid involved in forced convection can constantly be replaced and treated mathematically while natural convection cannot be reproduced or duplicated. Because natural convention is not consistent, it is known to complicate electrode process and therefore should be eliminated. This can be successfully reached by carrying out electrolysis in a thermostat-controlled system, in the absence of vibration.

References

1. C.A. Martinson, G. Van Schoor, K.R. Uren, D. Bessarabov, Characterisation of a PEM electrolyser using the current interrupt method. Int. J. Hydrog. Energy **39**(36), 20865–20878 (2014)
2. K. Watts, A. Baker, T. Wirth, Electrochemical synthesis in microreactors. J. Flow Chem. **4**(1), 2–11 (2014)
3. D. Pletcher, R.A. Green, R.C. Brown, Flow electrolysis cells for the synthetic organic chemistry laboratory. Chem. Rev. **118**(9), 4573–4591 (2017)
4. EIS02, A.A.N., *Electrochemical Impedance Spectroscopy (EIS) Part 2–Experimental Setup* (2011)
5. H.L. Oon, *A Simple Electric Cell, Chemistry Expression: An Inquiry Approach* (Panpac Education Pte Ltd, Singapore, 2007), p. 236. ISBN 978-981-271-162-5
6. R. Holze, Daniell cell, in *Electrochemical dictionary*, 2nd edn., ed. by A.J. Bard, G. Inzelt, F. Scholz (Springer, Berlin, 2012), pp. 188–189
7. G. Inzelt, Crossing the bridge between thermodynamics and electrochemistry. From the potential of the cell reaction to the electrode potential. ChemTexts, **1**(1), 2 (2015)
8. N. Elgrishi, K.J. Rountree, B.D. McCarthy, E.S. Rountree, T.T. Eisenhart, J.L. Dempsey, A practical beginner's guide to cyclic voltammetry. J. Chem. Educ. **95**(2), 197–206 (2017)
9. R.C. Weast, *CRC Handbook of Chemistry, and Physics*, 70th ed. (CRC Press, Boca Raton, FL, 1989), p. D-221
10. J.R. Swierk, *Electron transfer kinetics in water-splitting dye-sensitized photoelectrochemical cells* (The Pennsylvania State University, 2014)
11. T. Sun, D. Wang, M.V. Mirkin, Electrochemistry at a single nanoparticle: from bipolar regime to tunnelling. Faraday Discuss. **210**, 173–188 (2018)
12. K.B. Oldham, A Gouy–Chapman–Stern model of the double layer at a (metal)/(ionic liquid) interface. J. Electroanal. Chem. **613**(2), 131–138 (2008)
13. A.J. Bard, L.R. Faulkner, *Electrochemical Methods—Fundamentals and Applications* (Wiley, New York, 2001), p. 791
14. C. Kittel, *Thermal physics*, 3rd edn. (Wiley, New York, 1969), p. 215
15. S.A. Mamuru, *Electrochemical and Electrocatalytic Properties of Iron (II) and Cobalt (II) Phthalocyanine Complexes Integrated with Multi-Walled Carbon Nanotubes* (Doctoral dissertation, University of Pretoria, 2010)

Chapter 5
Properties and Synthesis of Metal Oxide Nanoparticles in Electrochemistry

Ikenna Chibuzor Emeji, Onoyivwe Monday Ama, Uyiosa Osagie Aigbe, Khotso Khoele, Peter Ogbemudia Osifo, and Suprakas Sinha Ray

Abstract The synthesis and study of "metal oxide nanoparticles", has gain greater attention over the past 10 years among interdisciplinary researchers. The major interest may be as a result of their unique physical and chemical properties, which gives rise to their various industrial usage in the field of catalysis, electronics, solar energy conversion, and others. As the particle size diminishes, the ratio of surface atoms to those inherent rises, enabling the surface properties to dictate the overall properties of the nano-materials. Also, metal oxide nanoparticles manifest different optical and electrical properties in proportion to that of the bulk material. Hence, as the size of the solid becomes smaller, the band gap becomes larger. This, therefore, gave scientists the unique opportunity of nanofabrication synthesizing highly complex nanostructure with different electronic and optical properties just by manipulating its particle size.

I. C. Emeji (✉) · O. M. Ama · P. O. Osifo
Department of Chemical Engineering, Vaal University of Technology, Private Mail Bag X021, Vanderbijlpark 1900, South Africa
e-mail: emejiiyk@gmail.com

O. M. Ama · S. S. Ray
Department of Chemical Science, University of Johannesburg, Doornfontein 2028, South Africa

DST/CSIR National Centre for Nano-Structured Materials, Council for Scientific and Industrial Research, Pretoria 0001, South Africa

U. O. Aigbe
Department of Physics, College of Science, Engineering and Technology, University of South Africa, Pretoria 0001, South Africa

K. Khoele
Department of Chemical, Metallurgy and Material Engineering, Tshwane University of Technology, Pretoria, South Africa

© Springer Nature Switzerland AG 2020
O. M. Ama and S. S. Ray (eds.), *Nanostructured Metal-Oxide Electrode Materials for Water Purification*, Engineering Materials,
https://doi.org/10.1007/978-3-030-43346-8_5

5.1 Introduction

Metal oxides are metals bonded with oxygen [1]. Different metallic groups such as alkali metals, alkaline-earth metals, etc. on reaction with oxygen, form crystalline solids that are made up of metal cation and an oxide anion. These oxides normally react with water to form bases or with acids to form salts. According to Sui and Charpentier [2], the most commonly used metal oxides in a wider variety of reactions are ZnO, MgO, BaO, Al_2O_3, SiO_2, PbO, TiO_2, ZrO_2, MoO_2, CuO, and V_2O_5. However, it has been widely reported that metal oxides have numerous possible application in chemistry, physics and material sciences [3, 4]. In the emerging field of nanotechnology, synthesized metal oxide nanoparticles, are found to possess a special kind of nanostructure [5, 6]. Hence, when metal oxides are brought into the nanometer scale, they exhibit improved physicochemical properties comparable to their bulk materials [7] as a result of demised size and elevated density of edge surface. Bulk oxides are specifically known for their unique robustness and usual stable systems with précised crystallographic structures. But adjustment in thermodynamic stability of the nanoparticles which is related to size can produce formation transformation and changes of cell parameters [8]. In order to display good structural stability and high mechanical strength, a nanoparticle must have a low surface free energy. These structural perturbances have been reported to be detected in several metal oxide nanoparticles. Nanoparticles are therefore defined as manufactured or natural materials which contains unbounded particles that has one of its dimension less than 100 nm [9]. They are either crystalline, amorphous or multi-crystal solids with loose particles [10]. They have uniform or irregular surface variations. In electrochemistry, the usage of metal oxide nanoparticles extends from the fabrication of chemical sensor [11], photocatalysts [12], microelectronic circuits and coatings of surfaces against corrosion, to energy conversion devices which include fuel and solar cells [13]. Their behaviour can be metallic, semiconductor or insulator depending on their structural geometries and inherent electronic structure.

5.2 Classification of Nanoparticles

Nanoparticles are usually classified based on their material composition into organic, inorganic and carbon-based.

5.2.1 Carbon-Based Nanoparticles

These are Nano-material completely made of carbon [14] and they are classified into Fullerenes (C60), graphene (Gr), carbon nanotubes (CNT), carbon nanofibers and carbon black as presented in Fig. 5.1. Except for Carbon black, these carbon-based

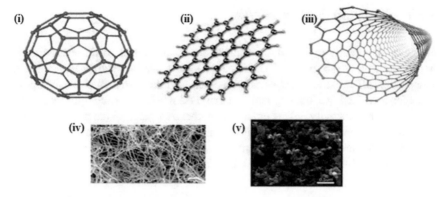

Fig. 5.1 Classes of carbon-based nanoparticles: (i) Fullerenes, (ii) Graphene, (iii) Carbon nanotubes, (iv) Carbon nanofibers and (v) Carbon black [14]

materials can be synthesized by Laser ablation, arc discharge, and chemical vapor deposition (CVD) methods [15].

5.2.2 Organic Nanoparticles

These are biologically inspired nanoparticles. In other words, they are organic polymers made from organic matter and they are non-toxic, eco-friendly and biodegradable. As suggested by Tiwari et al. [16] and illustrated in Fig. 5.2, examples of organic nano-materials are dendrimers, micelles, and liposomes. Observing their diagram as shown in Fig. 5.2, micelles and liposomes possess a hollow–cord called nanocapsules, which are responsive to radiation. Generally, organic nanoparticles are extensively used for biomedical applications especially as drug delivery carriers to a specific part of the body.

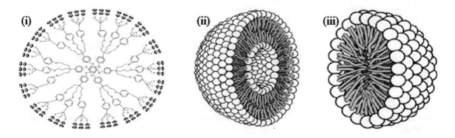

Fig. 5.2 Organic nanoparticles includes (i) Dendrimers, (ii) Liposomes and (iii) micelles [16]

5.2.3 Inorganic Nanoparticles

Inorganic nanoparticles include a range of metal and metal oxide nanoparticles. According to Salavati et al. [17], almost all metals can be synthesized into their nanoparticles. The oxide of metals nanoparticles is synthesized to improve their properties for increased productivity and reactivity. As when compared with their metal equivalent, the reactivity enhancement of synthesized metal oxide nanoparticles is with an exceptional property. The most synthesized oxide of metal nanoparticles is titanium dioxide (TiO_2), Aluminium oxide (Al_2O_3), Cerium oxide (CeO_2), Iron oxide (Fe_2O_3), Silicon dioxide (SiO_2), Zinc oxide (ZnO), Magnetite (Fe_3O_4).

5.3 Synthesis of Metal Oxide Nanoparticles

It has been reported that the microstructural, physicochemical properties of nanoparticles are controlled by their methods of synthesis.

As a result, numerous synthesis methods were used to synthesize metal oxide nanoparticles to enhance their properties and structures for a variety of industrial usage.

These methods are categorized into top-down and bottom-up methods [18] as illustrated in Table 5.1. While destructive techniques reduce bulk/larger materials to nanoparticles remains, constructive methods "grows" nanoparticles materials from simple molecules or atoms. Processes such as using micelles template to growth, functionalization of nanoparticles surface or limiting the surfactant concentrations may be used for nanoparticles size and shape control. Christian et al. [18] suggested that constructive methods used to synthesize nanoparticles depend on the law of supersaturation to control particle size as illustrated in Fig. 5.3. According to him,

Table 5.1 Different categories of synthesized nanoparticles [18]

Category	Methods	Nanoparticles
Bottom-up (constructive) methods	Sol-gel Electrospinning Chemical vapour disposition (CVD) Pyrolysis Biosynthesis	Carbon, metal and metal oxide based Organic polymers Carbon and metal based Carbon and metal oxide based Organic polymers and metal based
Top-down (destructive) methods	Mechanical attrition Lithography Laser ablation Sputtering Thermal decomposition	Clay, coal and metal powders Metal-based Carbon-based like carbon nanotubes; originates from metal oxide Originates from metal-based Carbon and metal oxide based

Fig. 5.3 A diagram showing a summary of nanoparticles formation [18]

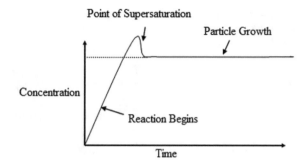

the concentration of the material increases with no precipitate until the limit of saturation is reached. Further reaction exceeding the limit of saturation resulted in precipitation from solution.

The formation of new material was noticed at the surface of the initial particle, where the growth of the final nanoparticles occurs. Hence, the surface chemistry and properties of the synthesized nanoparticles are as a result of the reaction medium. As shown in Table 5.1, the methods commonly used to prepare metal oxide nanoparticles are Sol-gel and flame pyrolysis methods.

5.3.1 Sol-Gel Method

The sol-gel process involves the conversion of precursor solution (usually metal salts or metal alkoxide) into a nano-structured inorganic solid through inorganic polymerization reactions catalyzed by water. The reaction involved is hydrolysis and condensation. Hydrolysis process is the aqueous dissolution of the metal salt or metal alkoxide in an organic solvent like alcohol, miscible in water to yield a homogeneous solution either by shaking, stirring or sonication. The homogeneous solution is then converted into Sol (A firm distribution of colloidal particles). A colloidal system is a well distributed processes, comprising of dispersion medium and a dispersion phase, which can either be a gas, liquid or gas. The liquid in liquid (emulsion), solid in gas (aerosols) and solid in liquid (emulsions) are all accepted phenomena's. Nature gives water and air as a dispersion medium, while inorganic solids or organics or their mixtures are regarded as dispersed phase media [19].

However, during hydrolysis, the OH (or O-R group) is being restored by the nucleophilic hostility of oxygen atom in the water molecules resulting to alcohol discharge and metal hydroxide (sol) formation. Condensation process gives rise to "Gel" development. The colloidal solution remains so as to mature. During maturity, condensational reaction occurs as metal hydroxide/alkoxy species bonds to release H_2O/R-OH. The resultant nanoparticles are recovered by sedimentation, drying, filtration and centrifugation [20]. Akiba [21], demonstrated the Sol-gel method by using

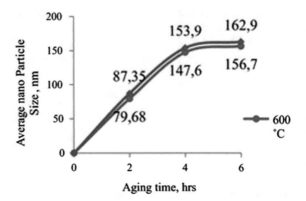

Fig. 5.4 Average nanoparticle size of SiO₂ powder [22]

the process to prepared fine iron oxide, hematite (α-Fe_2O_3) nanoparticles using Ferric Chloride as the precursor, ethanol as the solvent and ammonium hydroxide as the reducing agent. Initially, ferric chloride was dissolved in the solvent and the reducing agent was added drop by drop to produce the iron oxide nanoparticles. Akiba [21], XRD report, shows that the synthesized Nano-sized nanoparticles was in α-Fe_2O_3 phase exhibiting rhombohedra structure, thus indicating a high degree of purity of rhombohedra α-Fe_2O_3. Also work done by Azlina and co-workers [22] successfully synthesized silica nanostructures using sol-gel process, in which tetraethyl orthosilicate was used as a precursor, acetic acid as the catalyst and deionized water as the electrolyte. From their report, as indicated in Fig. 5.4, 2 h of calcination and heating at 700 °C yields relatively enlarged average particle size of 87.35 nm than those specimen calcined at 600 °C which has particle size of 79.68 nm. Major advantages of sol-gel techniques are inherent in their high yield, low operating temperatures and low production costs [23].

5.3.2 Pyrolysis Method

Pyrolysis is another bottom-up method used in the industries for large scale production of metal nanoparticles. It involves direct combustion of a precursor with flame. The precursor used in this process is either liquid or vapor and it is fed into the pyrolysis chamber or furnace consisting of high-pressure gas assisted nozzle from where it burns [24]. As the dispersed liquid burns in the flame, it evaporates and became converted to nanoparticles [25]. Nemade and Waghuley [26], uses solvent mixed spray pyrolysis technique to synthesize MgO nanoparticles with the excellent crystalline structure using magnesium nitrate ($Mg(NO_3)_2 \cdot 6H_2O$) and hexamethylenetetramine ($C_6H_{12}N_4$) liquid mixture. He further reported that the synthesized MgO nanoparticles have an average particle size of 9 nm. Because of the type of feedstock used, the pyrolysis process is a continuous process with high yield. Other major advantages of the process are inherent in its simplicity, cost-effectiveness and maximum productivity.

5.3.3 Others

Generally, other methods which have been developed, and broadly used to synthesize nanoparticles were indicative. Emulsion polymerization was reportedly used by Cho et al. [27] to synthesize polymeric nanoparticles. The spinning method was used by Tai and Co-workers [28] in a spinning disk reactor to synthesize polyhedral nanoparticles of 30–70 nm magnesium oxide nanoparticles. Chemical Vapour Deposition (CVD) method was used by Bhaviripudi et al. [14], to demonstrate the growth of Single-walled carbon nanotubes (SWNTs) using gold nanoparticle catalysts prepared by block copolymer templating technique. It was however observed from these methods of synthesis that changes in particle size distribution, shape, and phase influences the characteristic of the synthesized. Therefore, different methods of synthesis will have an effect on the purity, size allotment, surface modification and any fall-outs in the final nano-materials. Obviously, the techniques used to synthesize nanoparticles definitely have an enormous impact on its performance in any given application. Getting to know the effect of different factors during synthesis is paramount to understanding the synthesized Nano-material and results obtained in working with these materials.

5.4 Properties of Nanoparticles

5.4.1 Optical Properties

The properties of Nanomaterials (NMs) are the size and sharp dependent. In order to understand the physicochemical properties of NMs, one had to consider it as a tiny semiconducting nanoparticle called Quantum dots (QDs) with so small dimensions such that their diameters are 10 nm or less. However, a bulk semiconductor contains numerous atom which lead to the generation of orbital with different energy levels. Trindade et al. [29] reported that at 0 K the lower energy levels also called the valence band is completely filled with electrons, while the higher energy level which was labeled the conduction band is empty and unoccupied. The energy band gap (E_g), is the demarcation between the two bands as shown in Fig. 5.5, and its magnitude is an attribute of the bulk material. Therefore, the energy spectrum of a bulk semiconductor appears to be continuous as electrons are free to move around. On another development, an electromagnetic wave, allows excited electrons to migrate pass the band gap to the conduction band creating a hole in the valancy band.

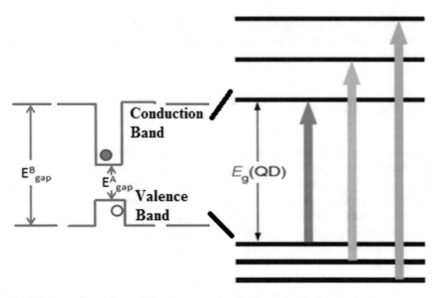

Fig. 5.5 Energy band diagram showing energy levels of a semiconductor

However, an excited electron in the conduction band and the resulting hole in the valence band form an "electron-hole pair". However, in a quantum-confined system, the material is reduced to a nanometer size. Confinement in this sense means restriction of electron movement to a specific energy level leading to a transition from continuous to discrete energy levels. Also, in reducing particle size to nano-size, the electron in the conduction band become restricted to a discrete energy level. Thus, as the size of a particle decreases to Nano-size, the electron in the conduction band or hole in the valance band become confined to a discrete energy level, thereby widening up the band gap and ultimately enlarging the band gap energy. Therefore, the quantum confinement effect corresponding to Nanoparticle size can be predicted using concepts such as De Broglie wavelength and exciton Bohr radius. Electron, hole, and exciton Bohr radius are given by (5.1), (5.2) and (5.3). So strong quantum confinement is observed when the nanoparticle radius r, is smaller than a_e, a_h and a_B.

$$a_e = \frac{0.0529\varepsilon}{m_e^*/m_0}\,(\text{nm}) \tag{5.1}$$

$$a_h = \frac{0.0529\varepsilon}{m_h^*/m_0}\,(\text{nm}) \tag{5.2}$$

$$a_B = \frac{0.0529\varepsilon}{\mu^*/m_0}\,(\text{nm}) \tag{5.3}$$

When using de Broglie wavelength to predict the chemical properties of the nanomaterials, Electron, hole and Broglie wavelength given in (5.4), (5.5) and (5.6) are used. Using this concept, quantum confinement effect was felt once the nano-particles properties such as diameter and wavelength of the electron became equal [30]

$$\lambda_e = \frac{h}{p_e} = \frac{h}{\sqrt{2m_e^* K_B T}} \tag{5.4}$$

$$\lambda_h = \frac{h}{p_h} = \frac{h}{\sqrt{2m_h^* K_B T}} \tag{5.5}$$

$$d = \frac{h}{\sqrt{2m_{e,h}^* K_B T}} \tag{5.6}$$

Due to quantum-size confinement, absorption of light becomes both discrete-like and size-dependent. Effective mass approximation (EMA) is another acceptable theoretical method used to explain the size dependence of the optical properties of nanometer semiconductors. The EMA theory predicts a r^{-2} dependence, with a main r^{-1} correction term in the strong electron-hole confinement regime. As explained herein, Li et al. [31], used Elliott's exciton absorption theory to report that undoped ZnO and Aluminium doped (AZO) films show identical quantum size dependence as a result of quantum confinement effect. On comparison with undoped ZnO films, he explained that doped AZO films exhibit an increase in the band gap energy and a decrease in the exciton binding energies because of free electron screening effect, which overpowers the excitonic absorption and enables the observation of blue shift of the absorption edge. In the same line, Viswanatha and co-workers in [32], were able to synthesize high quality ZnO nanocrystals with sharp absorption edges in four different sizes. The Bandgap energy obtained exhibits good agreement over the entire range of sizes but are in contrast, with the result obtained from theoretical EMA method.

5.4.2 Electrical Properties

Metal oxide nanoparticles have been described to exhibit varied ionic or mixed ionic/electrical conductivity depending on their sizes. Based on this variation, they are used for different industrial application by controlling its size. The effective amount of electronic charge transporter in any metal oxide is a function of the band gap energy according to the Boltzmann statistics. However, Trindade et al. [29] reported that conductivity (σ) in metal oxide nanoparticles is controlled by the amount of electron-hole pairs, which is the concentration of charge transporter (n) normally expressed in terms of the number of particles per cubic centimeter. Generally, electronic conduction can be referred to as n- or p-hopping-type depending on whether the principal charge carrier is either electron or hole. Enhancement of free electron-hole is achieved by introducing non-stoichiometry oxygen vacancies.

Table 5.2 Properties of selected metal oxide nanoparticles and their application

Metal oxide nanoparticles	Properties and application
TiO_2	Excellent magnetic properties with high surface area, In industries they are used for catalysis, photocatalysis and for pollutant elimination
Zinc oxide	ZnO are used as UV blockers in sun locations, mixed varistors, solar cell and optoelectronics, in gas sensors and also in catalysts for various types of organic reactions
MgO	The nanoparticles are of great important in the chemical industries where they are used as air scrubber and as a catalyst
Al_2O_3	Very sensitive to moisture, heat, and sunlight, highly reactive. They are used as support of active species in catalysis
CeO_2	They are used in catalysis, electrochemistry and material chemistry

Properties of selected oxide of metals and some of their industrial usage are shown in Table 5.2.

5.4.3 Particle Mobility

The electrical conductivity is associated with electron/particle mobility. Gravitational forces, buoyancy, and Brownian motion are factors controlling particle diffusion as stipulated by Einstein's law of diffusion (5.7). The frictional coefficient of the nanoparticle is related to the viscosity of the medium by Stokes law (5.8).

$$DF = kT \tag{5.7}$$

$$F = 6\pi \, \eta \, a \tag{5.8}$$

where D = diffusion coefficient, k = Boltzmann constant and T = temperature, F = the frictional coefficient for the particle, η = dynamic viscosity and a = particle radius.

This means that diffusion coefficient is inversely proportional to the particle size the mean displacement of a particle in time, t will be proportional to the reciprocal of the square root of the particle radius (r) and the mean displacement of a particle in time, t will be proportional to the reciprocal of the square root of the particle radius. Hence the smaller is the nanoparticle, the greater is the viscosity. It has been reported that nanoparticles have a very high superficial area to volume ratio which scopes with the inverse of the radius. This means that the number of surface atoms available to catalyze a chemical reaction corresponds inversely to the radius of the particle. Because of mentioned surface dissimilarity in the lattice structure of a particles, bulk materials may react in a different manner as nanoparticles. This is the reason why Yoo [33] reported that materials developed in the form of a nanoparticle posseses higher activity in catalytic processes than their bulk equivalent.

5.5   Conclusion

Nanotechnology has benefitted us daily by improving the achievement and effectiveness of everyday objects. Nanoparticles are tiny materials having particles of size ranging from 1 to 100 nm, with one or more dimensions. However, the general classification of nanoparticles into organic, inorganic and carbon-based particles in nanometric scale has produced better material as compared to bulk materials. Enhanced properties shown by nanoparticles include a high surface area to volume, stability, strength, improved sensitivity, high reactivity, etc. because of their diminished size structure. Nanoparticles are synthesized by numerous methods for industrial and domestic usage which are classified into physical, chemical and mechanical processes.

References

1. C.N.R. Rao, Transition metal oxides. Annu. Rev. Phys. Chem. **40**(1), 291–326 (1989)
2. R. Sui, P. Charpentier, Synthesis of metal oxide nanostructures by direct sol-gel chemistry in supercritical fluids. Chem. Rev. **112**(6), 3057–3082 (2012)
3. H.H. Kung, *Transition Metal Oxides: Surface Chemistry and Catalysis* (Elsevier, Amsterdam, 1989)
4. C. Noguera, *Physics and Chemistry at Oxide Surfaces* (Cambridge University Press, Cambridge, UK, 1996)
5. J.A. Rodriguez, G. Liu, T. Jirsak, Hrbek, Z. Chang, J. Dvorak, A. Maiti, J. Am. Chem. Soc. **124**, 5247 (2002)
6. M. Valden, X. Lai, D.W. Goodman, Onset of catalytic activity of gold clusters on titania with the appearance of nonmetallic properties. Science **281**(5383), 1647–1650 (1998)
7. H. Gleiter, Nanostructured materials, state of the art and perspectives. Nanostruct. Mater. **6**, 3–14 ((1995))
8. H. Zhang, J.F. Bandfield, J. Mater. Chem. **8**, 2073 (1998)
9. J. Jeevanandam, A. Barhoum, Y.S. Chan, A. Dufresne, M.K. Danquah, Review on nanoparticles and nanostructured materials: history, sources, toxicity and regulations. Beilstein J. Nanotechnol. **9**(1), 1050–1074 (2018)
10. S. Machado, J.G. Pacheco, H.P.A. Nouws, J.T. Albergaria, C. Delerue-Matos, Characterization of green zero-valent iron nanoparticles produced with tree leaf extracts. Sci. Total Environ. **533**, 76–81 (2015)
11. S.S. Weissenrieder, J. Müller, Thin Solid Films **300**, 30 (1997)
12. M. Anpo, K. Chiba, M. Tomonari, S. Coluccia, M. Che, M.A. Fox, Bull. Chem. Soc. Jpn. **64**, 543 (1991)
13. T. Yoshida, K. Terada, D. Schlettwein, T. Oekermann, T. Sugi-ura, H. Minoura, Adv. Mater. **12**, 1214 (2000)
14. S. Bhaviripudi, E. Mile, S.A. Steiner, A.T. Zare, M.S. Dresselhaus, A.M. Belcher, J. Kong, CVD synthesis of single-walled carbon nanotubes from gold nanoparticle catalysts. J. Am. Chem. Soc. **129**(6), 1516–1517 (2007)

15. N. Kumar, S. Kumbhat, *Carbon-Based Nanomaterials. Essentials in Nanoscience and Nanotechnology* (Wiley, Hoboken, 2016), pp. 189–236
16. D.K. Tiwari, J. Behari, P. Sen, Application of nanoparticles in waste water treatment. World Appl. Sci. J. **3**(3), 417–433 (2008)
17. M. Salavati-niasari, F. Davar, N. Mir, Synthesis and characterization of metallic copper nanoparticles via thermal decomposition. Polyhedron **27**, 3514–3518 (2008)
18. P. Christian, F. Von der Kammer, M. Baalousha, Th Hofmann, Nanoparticles: structure, properties, preparation and behaviour in environmental media. Ecotoxicology **17**, 326–343 (2008)
19. J.F. McCarthy, L.D. McKay, Colloid transport in the subsurface: past, present, and future challenges. Vadose Zone J. **3**, 326–337 (2004)
20. S. Mann, S.L. Burkett, S.A. Davis, C.E. Fowler, N.H. Mendelson, S.D. Sims, D. Walsh, N.T. Whilton, Sol-gel synthesis of organized matter . Chem. Mater. **9**(11), 2300–2310 (1997)
21. J.D.H. Akiba, Structural study on iron oxide nanoparticles prepared by sol-gel method. Int. J. Sci. Eng. Res. **9**(7), (2018)
22. H.N. Azlina, J.N. Hasnidawani, H. Norita, S.N. Surip, Synthesis of SiO_2 nanostructures using sol-gel method. Acta Phys. Pol., A **129**(4), 842–844 (2016)
23. C.H. Bartholomew, R.J. Farrauto, *Fundamentals of Industrial Catalytic Processes*, 2nd edn. (Wiley-Interscience, Hoboken, 2006)
24. H.K. Kammler, L. Mädler, S.E. Pratsinis, Flame synthesis of nanoparticles. Chem. Eng. Technol.: Ind. Chem.-Plant Equip.-Process Eng.-Biotechnol. **24**(6), 583–596 (2001)
25. W.Y. Teoh, R. Amal, L. Mädler, Flame spray pyrolysis: an enabling technology for nanoparticles design and fabrication. Nanoscale **2**, 1324–1347 (2010)
26. K.R. Nemade, S.A. Waghuley, Synthesis of MgO nanoparticles by solvent mixed spray pyrolysis technique for optical investigation. Int. J. Metals (2014), Article ID 389416, 4 p
27. Y.S Cho, S. Ji, Y.S. Kim, Synthesis of polymeric Nanoparticles by emulsion polymerization for particle self-assembly applications. J. Nanosci. Nanotechnol. **19**(10), 6398–6407 (2019)
28. C.Y. Tai, C.T. Tai, M.H. Chang, H.S. Liu, Synthesis of magnesium hydroxide and oxide nanoparticles using a spinning disk reactor. Ind. Eng. Chem. Res. **46**(17), 5536–5541 (2007)
29. T. Trindade, P. O'Brien, N.L. Pickett, Nanocrystalline semiconductors: synthesis, properties, and perspectives. Chem. Mater. **13**(11), 3843–3858 (2001)
30. J. Khatei, Semiconductor nanocrystals or quantum dots. Resonance 77 (2013)
31. X.D. Li, T.P. Chen, P. Li, Y. Liu, K.C. Leong, Effects of free electrons and quantum confinement in ultrathin ZnO flims: a comparison between undoped and Al-doped ZnO. Opt. Express. **21**(12), 14131–14138 (2013)
32. R. Viswanatha, S. Sapra, B. Satpati, P.V. Satyam, B.N. Dev, D.D. Sarma, Understanding the quantum size effects in ZnO nanocrystals. J. Mater. Chem. **14**(4), 661–668 (2004)
33. J.S. Yoo, Selective gas-phase oxidation at oxide nanoparticles on microporous materials. Catal. Today **41**, 409–432 (1998)

Chapter 6
Metal Oxide Nanomaterials for Biosensor Application

Azeez Olayiwola Idris, Onoyivwe Monday Ama, Suprakas Sinha Ray, and Peter Ogbemudia Osifo

Abstract Metal oxide nanomaterials (MONMs) have captivated a lot of attention due to their unique analytical properties such as fast electron transfers kinetics, large surface area, catalytic ability, conductivity (enhancement of charge flow), biocompatibility and excellent electrochemical properties. These properties have allured scientists towards employing MONMs as an upcoming platform for the development of various biosensors for different electroanalyte, biological compounds, biological molecules or cancer biomarkers. It is important to highlight that the early detection of cancer biomarkers will assist in solving the long-standing health problems in the world. In view of this, the present chapter underscores the construction of different biosensors using metal oxide nanomaterials as a novel platform for the determination of various biological molecules.

6.1 Introduction

Metal oxide nanomaterials (MoNMs) have fascinated so much interest due to their interesting analytical merits, which include high surface reactivity, chemical inertness, high surface area, fast electron transport and biocompatibility features [1, 2].

A. O. Idris (✉)
Nanotechnology and Water Sustainability Research Unit, College of Science, Engineering and Technology, University of South Africa, P/Bag X6, Florida Science Campus, Florida Park, 1710 Rooderpoort, Johannesburg, South Africa
e-mail: idrisalone4real@gmail.com

O. M. Ama · S. S. Ray
Department of Chemical Science, University of Johannesburg, Doornfontein, 2028, Johannesburg, South Africa

DST-CSIR National Center for Nanostructured Materials, Council for Scientific and Industrial Research, Pretoria 0001, South Africa

O. M. Ama · P. O. Osifo
Department of Chemical Engineering, Vaal University of Technology, Private Mail Bag X021, Vanderbijlpark 1900, South Africa

© Springer Nature Switzerland AG 2020   97
O. M. Ama and S. S. Ray (eds.), *Nanostructured Metal-Oxide Electrode Materials for Water Purification*, Engineering Materials,
https://doi.org/10.1007/978-3-030-43346-8_6

These properties attracted the use of MoNMs for the immobilisation of various biological molecules in biosensor development. MoNMs are sub-divided into organic, inorganic and organic-inorganic nanocomposites. Examples of inorganic nanostructured metal oxide nanomaterials are titanium (IV) oxide (TiO_2), zinc oxide (ZnO), manganese (IV) oxide (MnO_2), iron (III) oxide (Fe_2O_3) and zirconium (IV) oxide (ZrO_2). These metal oxide nanostructures have assisted in opening greater new doors of opportunities in different fields of research and science.

Interestingly, MoNMs have been reported to play significant roles in the immobilisation of biomolecules and facilitate electron communication between the working electrode and the electroanalyte. More so, modifying working electrode with MoNMs helped in enhancing the electrochemical performance of biosensors which include detection limits, selectivity, sensitivity and stability. However, they are very difficult to functionalise, but this problem is addressed by forming heterostructure and synergising them with organic nanostructures or dendrimer.

In the present work, this chapter discussed some of the published work on the application of metal oxide nanomaterials as smart nanomaterials for the construction of various biosensors.

6.2 TiO_2 Nanoparticles as a Smart Nanomaterial for the Construction of a Biosensor

Titanium (IV) oxide nanoparticles have received tremendous attention because of its unique morphology, analytical and physicochemical features such as high surface area, low cost, non-toxicity, physical and chemical stability [3]. These properties have fascinated the application of this metallic oxide for the fabrication of various biosensor. For instance, a glucose biosensor was developed by using titanium nanorod synthesised by facile method and glucose oxidase (GOx) is attached to its surface [3]. Interestingly, the limit of detection (LOD) and the sensitivity of the glucose biosensor were 2.0×10^{-6} M and 23.2 mA M^{-1} cm^{-2} [3]. More so, a conductometric glucose biosensor was constructed using titanium dioxide-cellulose hybrid nanocomposite and GO_x was attached into the nanocomposite by physical absorption. The biosensor has a linear response of 1–10 mM [4]. In addition to this, titanium oxide prepared from a sol-gel method and an organic copolymer (poly(vinyl alcohol) grafting 4-vinylpyridine were used for the preparation of a biosensor. However, it was reported that the bioactivity of the biosensor was retained even after the glucose oxidase was immobilised on the nanocomposite. The following were deduced from the constructed biosensor: a rapid response time of 20 s, a concentration range of 9 mM and a sensitivity of 405 nA/mM were reported [5]. Furthermore, titanium oxide nanotube was prepared via hydrothermal synthesis and aniline was polymerised on the prepared titanium oxide nanotube by oxidative polymerisation to form PANI-TiO_2 nanotube composite for the immobilisation of GO_x in the development of an electrochemical biosensor, this platform displayed good sensitivity (11.4

μA mM^{-1}), wide linear range of 10–250 μM and detection limit of 0.5 μM [6]. In another report, an amperometric $C_6H_{12}O_6$ biosensor was designed by encapsulation of micro-sized TiO_2 particles on graphene nanosheets (GR) to form TiO_2-GR composites, the amperometric biosensor has a linear concentration range of 0 to 8 mM, applied potential and sensitivity of -0.6 V and 6.2 μA M^{-1} cm^{-2} [7]. Similarly, microwave irradiation was used in the synthesis of nanocluster titanium oxide alongside with rGO to form a rGO-TiO_2 hybrid, which was further employed for the entrapment of GOx and in the construction of biosensor for glucose. Interestingly, this glucose biosensor was reported to have a short response time of 10 s, low work potential of -0.7 V, wide linear range of 0.032–1.67 mM, detection limit of 4.8 μM, a sensitivity of 35.8 μA mM^{-1} cm^{-2} and small Michaelis-Menten constant K_m (0.81 mM) [8]. It is important to highlight that small K_m value is a reflection of the affinity of the biological element to its analyte.

Also, glucose oxidase was immobilised on titanium oxide alongside with graphene oxide for the electroreduction and electrooxidation of H_2O_2, this nanobiocomposite showed impressive photoelectrocatalytic activity towards H_2O_2 in the absence and presence of light [9]. Similarly, lactate oxidase was immobilised on titanium oxide/graphene nanocomposite in the development of an enzymatic electrochemical biosensor for lactate sensor via hydrogen peroxide detection [10]. More so, 3D graphene nanostacks alongside with titanium dioxide nanowire was used in the fabrication of cholesterol biosensor [11]. In addition to this, co-immobilisation of horseradish peroxidase (HRP) and chitosan were incorporated to Au-modified titanium oxide nanotubes arrays for the fabrication of novel H_2O_2 biosensor, a detection limit of 4×10^{-4} mol/L was reported [12]. In the same vein, gold nanoparticle was used to dop titanium oxide for the construction of H_2O_2 biosensor, in this case, HRP was not involved in the fabrication process, the biosensor exhibited a rapid electrocatalytic response of 3 s and a limit of detection of 0.2 μM was reported [13]. Sometimes, heterostructure are formed using different metal oxides in the fabrication of photoelectrochemical and electrochemical biosensor, for instance, synergistic combination of two metal oxides which are Cu_2O and TiO_2 nanotubes array were used for the fabrication of non-enzymatic biosensor for the quantification of hydrogen peroxide, the result revealed that the biosensor displayed good catalytic performance with rapid response time of 4 s, linear concentration range of 0.5–8 mM, detection limit, sensitivity and applied voltage of 90.5 μM, 179.9 μA mM^{-1} cm^{-2} and -0.4 V was reported [14]. In the same vein, glucose oxidase was immobilised on titanium (IV) oxide alongside with silicon (IV) oxide nanocomposite in the fabrication of glucose biosensor, a limit of detection 1.2×10^{-10} M was reported [15]. In addition to this, tungsten oxide and titanium oxide were coated on indium tin oxide electrode (ITO) in the development of a multi-functionalised biosensor for simultaneous of norepinephrine and riboflavin [16].

Furthermore, titanium (IV) oxide has played significant roles in the fabrication of photoelectrochemical biosensor because it has a wide band gap of 3.2 eV, but the rapid recombination of electron-hole pair had drastically impeded the application of titanium dioxide or in general, semiconductors for photocatalysis and photoelectrochemical biosensor. To improve the photoelectrochemical properties of titanium oxide, different materials including quantum dots, semiconductor coupling, dye sensitisers, metal and metal atoms doping have assisted in improving its photoresponse in the visible light region [17, 18]. For example, Cds quantum dot was used to enhance the photocurrent signal of titanium dioxide nanotube for the determination of asulam [19]. In addition to this, graphite-like carbon nitride (g-C_3N_4) was used to form a composite with TiO_2 nanosheets for the fabrication of photoelectrochemical glucose biosensor (PEC). It was reported that the (g-C_3N_4) assisted in minimising the band gap of the composite, helps the PEC biosensor to be excited in the visible region and prevent the deactivation of glucose oxidase by UV light [20]. Similarly, graphite-like carbon nitride alongside with TiO_2 was equally employed for the construction of photoelectrochemical glucose and lactose biosensor, but in this case, gold nanoparticle and manganese (IV) oxide nanoparticle were incorporated in the formation of the heterojunction, it was reported that the large surface area of MnO_2 nanoparticle assisted in immobilisation of the enzymes (glucose oxidase and β-galactosidase) while gold nanoparticle helped in improving photocurrent signal and PEC response in the measurements [18].

In another report, a self-powered visible light sensitive photoelectrochemical (PEC) glucose biosensor was designed from the formation of p-n heterojunction between Co_3O_4 and TiO_2 nanoparticles, although single-walled carbon nanotubes (CNTs) was also incorporated into this heterojunction so as to enhance the photo-conversion efficiency and photocurrent generation, the following were obtained from the PEC biosensor: linear concentration range 0–4 mM, sensitivity and LOD of 0.3 $\mu A \ mM^{-1} \ cm^{-2}$ and 0.16 μM were reported [21]. In the same vein, multi-walled CNTs and TiO_2 nanotube composite were used for the construction of a novel PEC lactic dehydrogenase (LDH) biosensor. It is important to highlight the analytical roles played by the MWCNT for the fabrication process. It helps to enhance the electrical conductivity of the semiconductor and improve visible light absorption, assisted in moving the photo-generated electrons away from the TiO_2 and helps in slowing down the recombination of holes and electrons [22]. Interestingly, this PEC biosensor was used in the regeneration of (NAD^+) in the enzymatic cycle and also in the immobilisation of the enzyme (lactic dehydrogenase) on the working electrode surface. The application of the (LDH) biosensor was interrogated by measuring the photocurrent [22].

Also, nitrogen-doped bio-based carbon (NDC) encapsulated on titanium (IV) oxide nanoparticle was synthesised via green synthetic route and subsequently functionalised with glucose oxidase on indium-tin-oxide (ITO) electrode for the construction of photoelectrochemical glucose biosensor [23]. It was documented that the sensitivity of the PEC biosensor was ascribed to the synergistic effects between the bio-derived nitrogen and carbon atoms in NDC, which increase the conductivity and also facilitate charge separation between holes and electrons [23]. Similarly, a simple one-pot 3D nitrogen-doped graphene hydrogel alongside with Ag and TiO_2 were used for the construction of a label-free photoelectrochemical thrombin aptasensor, the incorporation of metal nanoparticles and nitrogen-doped graphene assisted in facilitating charge separation efficiency and retard the recombination of electrons and holes [24]. Furthermore, a one-pot hydrothermal route was used to synthesise nitrogen-doped graphene silver-titanium (IV) oxide in the construction of PEC aptasensor for the determination of Pb^{2+}. It was reported that doping-nitrogen in graphene networks does not only help in speeding up charge transfer between the close carbon atom but also assist in enhancing photocurrent signal [25].

6.3   ZnO Nanoparticle as a Brilliant Nanomaterial for the Fabrication of a Biosensor

On the other hand, zinc oxide has attracted a lot of interest in recent years due to their analytical properties, which include large surface area, high electron communication, biocompatibility, wide band gap of 3.37 eV and n-type semiconductor for photo-conductivity and photocatalytic ability, large exaction binding energy of 60 meV, non-toxicity and excessive electrochemical stability [26–28]. The biocompatibility and large surface area of zinc oxide are the major properties that fascinated the application of ZnO for the development of an electrochemical biosensor, for instance, zinc oxide nanorod was used in the construction of $C_6H_{12}O_6$ biosensor, the detection was carried out in (PBS) solution within a concentration range of 1 μM–10 mM and human plasma [29]. The quantification of glucose on human blood plasma revealed the analytical significant and contributions of zinc oxide nanoparticle [29]. Similarly, zinc oxide nanoparticle synthesised by green synthetic route embedded with nitrogen-doped carbon sheet was employed in the design of a high-performance glucose biosensor, this biosensor was employed in the quantitative monitoring of $C_6H_{12}O_6$ in human blood serum [28].

In addition to this, zinc oxide in conjunction with glucose oxidase was used in the development of photoelectrochemical biosensor, it was reported that the structure of glucose oxidase was conserved after conjugation with zinc oxide nanoparticle, this must have resulted from the biocompatibility feature of this semiconductor [30]. More so, zinc oxide was employed in the construction of non-enzymatic biosensor, for example, Dayakar et al., reported a novel non-enzymatic glucose biosensor using zinc nanoparticle synthesised from leaf extract *Ocimium teniflorum*. The fabricated GCE/ZnO glucose biosensor possessed superior electrocatalytic activity, a detection limit of 0.043 μM, a rapid response time of 4 s and sensitivity of 631.30 μA mM^{-1} cm^{-2} were reported [31]. Furthermore, hydrothermal synthetic method was used for the preparation of n and p-type semiconductors (ZnO–CuO), this heterostructure was used for the fabrication of non-enzymatic glucose biosensor, it was reported that the uniform growth and stability of the nanocomposite were attributed to the presence of cationic polymer polyethyleneimine (PEI), which interact with the negatively charged heterostructure via electrostatic interaction and an impressive sensitivity of 1475.5 μA mM^{-1} cm^{-2}, 0.038 μM LOD and long term stability of 38 days were reported [32]. Similarly, a three-dimensional hybrid graphene in conjunction with nickel oxide and zinc oxide nanorod were used for the design of non-enzymatic $C_6H_{12}O_6$ biosensors, the fabricated biosensor was reported to possess the following features: wide linear range of 0.5 μM–1.11 mM, electrocatalytic response of less than 3 s, sensitivity and LOD of 2030 μA mM^{-1} cm^{-2} and 0.15 μM were reported [33]. The fast electrocatalytic response was accredited to the large surface area and high electroactivity of the fabricated biosensor prepared from the heterostructure. Interestingly, this biosensor was reported used in the quantification of $C_6H_{12}O_6$ in human serum samples [33].

6.4 MnO$_2$ Nanoparticle as an Intelligent Nanomaterial for the Development of a Biosensor

Another exciting metal oxide nanomaterial that has not been fully exploited in biosensor applications is manganese oxide (MO) nanoparticles. It has variable oxidation states which include (II, III, IV and VII), among the oxidation states of MO, MnO_2, Mn_3O_4 and $MnOOH$ are the most commonly employed for electrochemical applications [34]. The analytical properties of MO include high surface area, excellent biocompatibility, environment-friendliness, low cost and catalytic properties [35, 36]. These properties attracted scientists in using MO as a potential nanomaterial for the fabrication of different biosensors. For instance, we employed manganese (IV) oxide nanorod in conjunction with gold nanoparticle for the construction of biosensor for alpha-feto protein (AFP)—a cancer biomarker. In this report, alpha-feto protein was detected using electrochemical impedance spectroscopy and square wave voltammetry, detection limits of 0.00173 and 0.00276 ng/mL were obtained from both techniques [37]. The proposed equation for the manganese oxide nanorod synthesis is depicted in (6.1).

$$3MnSO_4(s) + 2H_2O(l) + 2KMnO_4(s) \rightarrow K_2SO_4(aq) + 2H_2SO_4(aq) + 5MnO_2(s)$$
$$(6.1)$$

In another report, a gold nanoparticle in connection with manganese oxide nanowire was utilised in the quantification of prostate-specific antigen (PSA). The immunosensor was prepared by growing gold nanoparticles on the screen-printed electrode (SPE) followed by the electrodeposition of MnO_2 nanowire on the AuNPs/SPE to form a 3D network with high surface area and glucose oxidase was wired as an enzyme label and 3,3′,5,5′-tetramethylbenzidine (TMB) as a redox terminator. The fabricated biosensor was used in the detection of PSA with a linear range of 0.005–100 ng/mL and 0.0012 ng/mL detection limit was reported [38]. Similarly, carbon nanotubes in connection with manganese dioxide (CNTs/MnO$_2$) composites were used as an electrochemical tag for electrochemical detection of alpha-feto protein—a cancer biomarker or a tumour marker for diagnosing hepatocellular carcinoma. The fabricated immunosensor displayed linear concentration range of 0.2–100 ng/mL and 40 pg/mL detection limit was reported [39].

More so, choline-based biosensor was constructed by depositing mixtures of solution containing (2.0 ml of 0.4% chitosan, 3.0 mg/mL MnO_2-NPS and 1.0 mg/mL choline oxidase (ChOx) on a GCE), the electrode was denoted as ChOX/MnO_2/Chit/GC also known as the biosensor and was further used in the detection of choline chloride and for the simultaneous oxidation and reduction of H_2O_2 [40]. It was reported that the simultaneous oxidation and reduction of H_2O_2 using MnO_2 nanoparticles is due to its bi-electrocatalytic property. It is important to note that choline is distributed in the CNS of mammals and a vital component of phospholipids, which is needed for the production of the neurotransmitter acetylcholine precursor [40].

Similarly, Guangxia et al. reported a facile choline-based biosensor using MnO_2 nanoparticles; the biosensor was developed by electrodepositing thin film MnO_2 on disk glassy working electrode using a mixture of (45 mM H_2SO_4 and 10 mM $KMnO_4$) at a deposition time of 200 s and a working potential of 0.2 V. After the electrodeposition process, the film was gently washed on the GCE with distilled water and allowed to dry for 600 s at ambient temperature. Hereafter, 5 μL of choline oxidase prepared in PBS (pH 8.0) was immobilised on the surface of MnO_2 modified electrode for 2 h and 2 μL of 0.5% Nafion was used as a binder to cover the ChOx/MnO_2 film on the surface of the electrode. The choline based-biosensor was employed in the electrochemical quantification of choline chloride using CV; a detection limit of 5 μM was reported. Interestingly, the fabricated biosensor displayed satisfactory results when used in the quantification of choline in real life samples [41].

Also, an amperometry quantification of H_2O_2 and $C_6H_{12}O_6$ were reported using CPE modified with Mn-based perovskites-type oxides, as a control experiment, the GOx was immobilised on the CPE with and without the Mn-based perovskites-type oxides for the quantification of $C_6H_{12}O_6$, it was reported that the sensitivity of the biosensor with or without the modifier were 158.1 and 11.9 μA mol^{-1}. This great improvement in the sensitivity revealed the analytical significance of the catalytic activity of the manganese-based perovskites-type oxides [42].

In another report, dual-function amperometric sensors were employed for the determination of H_2O_2 and $C_6H_{12}O_6$ using poly (diallydimethylammonium chloride) functionalised with rGO, MnO_2 and AuNPs nanocomposite. The synthesised nanocomposite was immobilised on the surface of the GCE for the construction of a non-enzymatic sensor for hydrogen peroxide, the results obtained revealed the impressive electrocatalytic ability towards H_2O_2 with sensitivity and detection limit of 1132.8 μA mM^{-1} cm^{-2} and 0.6 μM [43]. For the detection of glucose, the $C_6H_{12}O_6$ biosensor displayed remarkable performance with detection limit, sensitivity and small K_m value of 1.8 μM, 83.7 μA mM^{-1} cm^{-2} and 1.54 mM [43]. Similarly, a reagentless hydrogen peroxide was developed using methylene blue (MB) intercalated into the layer of MnO_2 (birmessite) to form (MB-Bir); the MB-Bir acts as the electron mediator. The amperometric measurements and cyclic voltammetry results revealed that MB co-immobilised with horseradish peroxidase (HRP) display good stability and the electron was efficiently shuttled between HRP and the GCE. The linear response of the biosensor was 1.5×10^{-6} M to 8.1×10^{-3} M and a detection limit of 1.3 μM was reported, the biosensor was able to retain 85% of its initial current response after 60 days storage and was selective in the presence of interference compounds which include sucrose, acetic acid, citric acid and L-ascorbate [44]. A non-enzymatic biosensor was developed using manganese (IV) oxide ultrathin nanosheets, the biosensor possessed high electrochemical ability for the detection of hydrogen peroxide in PBS using potential of +0.7 V, the following was deduced from the fabricated non-enzymatic biosensor: linear concentration range of (5×10^{-6} M to 3.5×10^{-3} M), high sensitivity of 130.56 μA mM^{-1} cm^{-2} and limit of detection 1.5 μM were calculated. The high sensitivity obtained for the biosensor was due to electrocatalytic properties and high surface to volume ratio of MnO_2 ultrathin nanosheets [45].

More so, a screen-printed amperometric biosensor was developed using MnO_2 nanoparticle as a mediator and glucose oxidase as a biocomponent for selective detection of $C_6H_{12}O_6$ in various compounds which include cellobiose, saccharose and beer samples using flow-injection analysis. The biosensor was reported to exhibit good reproducibility and stability [46]. Furthermore, MnO_2-NP was encapsulated with graphene nanoribbons (GNR) in the development of a biosensor for $C_6H_{12}O_6$, the biosensor was prepared by drop coating the synthesised MnO_2/GNR on the surface of SPE, followed by the immobilisation of glucose oxidase and Nafion was drop coated on its surface, the concentration range for the biosensor was from 0.1 to 1.4 mmol/L, LOD and sensitivity of 0.05 mmol/L and 56.32 μA mM^{-1} cm^{-2} were documented [47]. Interestingly, this biosensor was also employed in the electrochemical detection of H_2O_2, the proof of concept experiment revealed that when the biosensor was constructed using manganese (IV) oxide nanoparticles without the graphene nanoribbons, a detection limit of 0.23 mmol/L in comparison to 1.37 mmol/L obtained in using only graphene nanoribbons, this result further confirmed the excellent electrocatalytic ability of manganese (IV) oxide nanoparticle [47]. Furthermore, an amperometric glucose based-biosensor was designed using mixed-valence manganese oxides (cryptomelane-type and todorokite-type) on a carbon paste electrode; the operating potential was 0.3 V (vs. Ag/AgCl) and limit of detection 0.1 mmol/L was calculated [48]. The biosensor was reported to be stable and exhibited good reproducibility.

Li et al., employed manganese dioxide nanowire as an electrochemical platform in the construction of an enzymatic $C_6H_{12}O_6$ based-biosensor, the biosensor was prepared by adding a mixture of MnO_2 and 0.1% Nafion solution, the resulting solution was drop dried on a GCE at room temperature for 24 h, the electrode was assigned as (Naf/MnO_2/GCE), followed by drop drying 5 μL of enzyme solution on its surface, the enzyme solution was prepared by dissolving (15 mg/mL glucose oxidase in 0.1 M PBS (pH 7.0). Finally, glutaraldehyde was immobilised on the resulting electrode as a cross-linker. The linear concentration range for the glucose detection is from 0.2 to 3.8 mM; LOD of 25.56 μM and a sensitivity of 38.20 μA mM^{-1} cm^{-2} were calculated [49].

Moreover, polyaniline (PANI) and MnO_2 nanocomposites were co-electrodeposited on stainless steel working electrode by cycling a potential from -200 to 1100 mV for twenty cycles at a scan rate of 50 mV/s. Subsequently, the enzyme urease was immobilised on the surface to form PANI/MnO$_2$/Urease biosensor. The Chronoamperometric response of the fabricated biosensor revealed that the current signal increases as concentration increases. It was reported that the high sensitivity of the biosensor was due to the deposition of MnO$_2$ on the PANI film [50]. It is essential to highlight that few reports have been noted in the application of MnO$_2$ for the fabrication of photoelectrochemical biosensor in the literature. However, Tio et al., reported photoelectrochemical glucose and lactose biosensor prepared by decorating gold nanoparticles, manganese (IV) oxide nanoparticles and graphitic carbon nitride g-C$_3$N$_4$ on titanium (IV) oxide nanoparticles. It was reported that a p-n heterostructure was formed among the modifiers (MnO$_2$, g-C$_3$N$_4$ and TiO$_2$), this heterostructure prevent the recombination of excited electrons and promote charge transport on the working electrode. The linear measurement for the glucose detection was 0.004–1.75 mM, applied potential of 0 V and a sensitivity of 1.54 μA mM^{-1} cm^{-2}. For the lactose measurement, a sensitivity of 1.66 μA mM^{-1} cm^{-2}, applied potential of -0.4 V and linear concentration range of 0.008–2.5 Mm were reported [18].

6.5 Magnetic Oxide Nanomaterial as an Excellent Nanowire for the Design of an Electrochemical Biosensor

Magnetic oxide nanoparticle (MNP) is another excellent nanowire that is employed in the design of an electrochemical biosensor and allows the transfer of an electron between bulk electrode materials and redox systems due to their superb analytical features such as high conductivity, non-genotoxic, biocompatibility and a large surface to volume ratio [51]. The MNP can be immobilised on the surface of the transducers by chemical-covalent bonding, physical adsorption and electrodeposition for electrochemical amplification and signal generation [51]. Interestingly, iron-base oxide nanoparticles provide a large surface to volume ratio to encapsulate or immobilised various biochemical species or elements, thus, a low LOD is obtained [51]. Due to this aforementioned feature, iron oxide nanoparticles have been utilised in the assembly of various biosensors, for example, Rahman et al. were able to synthesise iron-oxide nanoparticles via green synthetic routes using Gooseberry leaves as a reducing agent. Although the average size of the nanoparticles was about 10 nm with some agglomeration, it was further used for the fabrication of glucose biosensor with 50% sensitivity lasting up to one day [52]. Furthermore, an electrochemical paper-based biosensor for carcinoembryonic antigen (cancer biomarker) was developed using iron oxide nanoparticles decorated on poly (3,4-ethylene dioxythiophene): poly (styrene sulphonate) (PEDOT:PSS). It was reported that the incorporation of Fe$_3$O$_4$ nanoparticles into PEDOT: PSS assisted in enhancing the signal stability and electrochemical performance of the biosensor. The proposed disposable paper-based biosensor exhibited impressive electrical conductivity, long-term stability, high heterogeneous electron transfer rate constant and high sensitivity [53]. Similarly, a surface plasmon resonance (SBR) biosensor was developed in the quantification of mouse IgG using gold nanorods-Fe_3O_4 nanohybrids. The nanohybrid was synthesised by taking advantage of the affinity between the negatively charged Fe_3O_4 and positively charged gold nanorods. Interestingly, the simplicity of the biosensor fabrication process was borne out of the ease in the immobilisation of the nano-hybrids on the biosensor surface using a magnetic pillar. The biosensor display was responsive in the concentration range of 1.25–40.0 μg m/L [54]. In another report, iron oxide nanoparticle (IONPs) was employed in the fabrication of glucose biosensor, for example, an amperometric glucose biosensor was constructed by immobilising (IONPs) alongside with GOx and Nafion on (SPCE). The IONP were synthesised via precipitation and was functionalised with citric acid (CA) to provide a functional group and hydrophilic surface for the immobilisation of glucose oxidase. The biosensor has LOD and sensitivity of 7 μM and 5.31 μA mM^{-1} cm^{-2} [55]. Similarly, ITO glass electrode was modified with IONPs alongside with Nafion for the development of $C_6H_{12}O_6$ biosensor, cyclic voltammetry was used in quantification of the different concentration of glucose with a sensitivity of 70.1 μA mM^{-1} cm^{-2} was reported [56].

More so, iron oxide nanoparticle was synthesised using co-precipitation method followed by the dissolution of chitosan solution (CH) to form CH-Fe_3O_4 nanocomposite, thereafter, GOx was immobilised on its surface by physical adsorption to form GOx/CH/Fe_3O_4/ITO nanobioelectrode. The following parameters were obtained from the fabricated biosensor: sensitivity of 9.3 μA/(mg dL cm^2), rapid response time of 5 s, shelf life of 8 weeks after refrigerating, linear concentration of 10–400 mg dL^{-1} glucose solution and Michaelis-Menten (K_m) value of 0.141 mM, the low value of K_m revealed the strong affinity of the GOx towards the analyte [57].

In another report, rGO/Fe_3O_4 was prepared in situ by electrodeposition of Fe_3O_4 on graphene sheets. Although graphene oxide (GO) was reduced to graphene by Fe^{2+}, thereafter, glucose oxidase was immobilised on the nanocomposite (rGO/Fe_3O_4) in the assembly of glucose biosensor. The constructed glucose biosensor (rGO/Fe_3O_4/GOx/GCE) displayed good electrocatalytic activity and fast electron transfer because of the excellent analytical properties of the rGO/Fe_3O_4 nanocomposite [58]. In addition, a reagentless glucose biosensor was constructed by immobilisation of GO_x to rGO/IONPs nanocomposite via electrostatic attraction on the SPE. The rGO-Fe_3O_4 nanocomposite was reported to have the following advantages: large surface area, excellent superparamagnetic property, favourable environmental conditions for enzyme immobilisation and facilitates electron communication between the enzymes and SPE. In addition to this, the IONPs help increases the surface area, assist in the immobilisation of the enzymes (GO_x), prevents the leaching of the enzymes on the electrode surface and assists in the stability of the biosensor. The following parameters were deduced from the reagentless glucose biosensor: amperometric rapid response of 3 s, a detection limit of 0.1 μM, deposition potential of − 0.45 V and sensitivity of 5.9 μA/mM. It is important to highlight that self-assembly and covalent bonding are the hallmarks for the mediatorless glucose biosensor [59].

A selective and sensitive $C_6H_{12}O_6$ biosensor was designed by deposition of Pt-NPs on Fe_3O_4/MWCNTS/Chitosan magnetic nanocomposite and used for the modification of GCE, thereafter, GO_x was immobilised on the surface. Each stage of the construction process was electrochemically characterised using CV, EIS and chronoamperometry. Interestingly, Fe_3O_4 was reported to play a pivotal role in the transfer of electron and strong support for Pt-NPs. The biosensor has long-term storage, good anti-interference ability and was used in the quantification of $C_6H_{12}O_6$ in serum sample [60].

In addition, IONPs were employed as a platform in the development of urea biosensor due to its unique large surface area and biocompatibility properties, which create a conducive or favourable environment for the immobilisation of enzymes. For instance, Atif et al. reported the application of IONP synthesised by co-precipitation method in the fabrication of potentiometric urea biosensor. The IONP was functionalised via electrostatic immobilisation of urease to the Fe_3O_4-chitosan nanobiocomposite, the biosensor was able to measure the different concentration of urea from 0.1 to 80 mM, sensitivity of 42 mV/decade and 12 s response [61]. Most importantly, the biosensor was able to monitor the concentration of urea in drugs, human serum and food industry-related samples [61]. Similarly, a biosensor was prepared by deposition of IONPs-chitosan-based nanobiocomposite film on (ITO) by physical adsorption followed by co-immobilisation of urease and glutamate dehydrogenase (GLDH). It was stated that the presence of IONPs helped in increasing the electroactive surface area of $(CH–Fe_3O_4)$ nanobiocomposite for the attachment of the enzymes, facilitates electron communication and increases the shelf life of the biosensor. The low value of $(K_m, 0.56$ mM) revealed the high affinity of the enzyme urease and glutamate dehydrogenase (GLDH) towards urea [61]. In another report, a urea biosensor was developed by modifying Au-electrode with poly(glycidyl methacrylate)-grafted iron oxide nanoparticle followed by the immobilisation of urease enzymes. It is important to highlight that IONPs were coated with poly(glycidyl methacrylate) in other to obtain a good enzyme immobilisation platform. The biosensor displayed rapid response time of 8 s, high sensitivity and wide linear range [62].

Generally, the integration of metal oxide nanoparticles with electrochemical transducers has played significant roles in point of care devices. They have played vital roles in the immobilisation of biomolecules and promote the transfer of an electron between the working electrode and analyte. Modifying electrode with MO-NPs has been reported to enhance the electroactive surface area, rapid mass transport, low detection limit, stability, sensitivity and selectivity.

6.6 Conclusions and Future Outlook

This chapter focuses on the applications of some selected metal oxide nanomaterials as a unique platform for the construction of various biosensor, the analytical properties of metal oxide nanoparticles such as large surface area, biocompatibility, conductivity and excellent electrochemical properties assisted in the detection of various biological molecules.

Unfortunately, few reports on the application of metal oxide nanoparticles for photoelectrochemical biosensors have been reported in the literature due to the wide bandgap possess by most metallic oxides, but the fast recombination of electron-hole pair had discouraged their application in a photoelectrochemical biosensor. This problem can be addressed by forming heterostructure with other metallic oxide nanoparticles. More work needs to be done in forming different heterostructure among various metallic oxides to increase their applications for photoelectrochemical biosensors. We also recommend the application of biosensor for multiple detections of various biological molecules as this will help to save time and cost.

References

1. S. Nizamuddin, M.T.H. Siddiqui, N.M. Mubarak, H.A. Baloch, E.C. Abdullah, S.A. Mazari, G.J. Griffin, M.P. Srinivasan, A. Tanksale, *Iron Oxide Nanomaterials for the Removal of Heavy Metals and Dyes From Wastewater* (Elsevier Inc., Amsterdam, 2019). https://doi.org/10.1016/b978-0-12-813926-4.00023-9
2. S.R.V.S. Prasanna, K. Balaji, S. Pandey, S. Rana, *Metal Oxide Based Nanomaterials and Their Polymer Nanocomposites* (Elsevier Inc., Amsterdam, 2019). https://doi.org/10.1016/b978-0-12-814615-6.00004-7
3. Z. Yang, Y. Tang, J. Li, Y. Zhang, X. Hu, Facile synthesis of tetragonal columnar-shaped TiO_2 nanorods for the construction of sensitive electrochemical glucose biosensor. Biosens. Bioelectron. **54**, 528–533 (2014). https://doi.org/10.1016/j.bios.2013.11.043
4. M. Maniruzzaman, S. Jang, J. Kim, Titanium dioxide—cellulose hybrid nanocomposite and its glucose biosensor application. Mater. Sci. Eng., B **177**, 844–848 (2012). https://doi.org/10.1016/j.mseb.2012.04.003
5. X. Chen, S. Dong, Sol-gel-derived titanium oxide/copolymer composite based glucose biosensor. Biosens. Bioelectron. **18**, 999–1004 (2003)
6. J. Zhu, X. Liu, X. Wang, X. Huo, R. Yan, Chemical Preparation of polyaniline—TiO_2 nanotube composite for the development of electrochemical biosensors. Sensors Actuators B. Chem. **221**, 450–457 (2015). https://doi.org/10.1016/j.snb.2015.06.131
7. H. Dong, S. Kyung, H. Chang, K. Roh, J. Choi, A glucose biosensor based on TiO_2—graphene composite. Biosens. Bioelectron. **38**, 184–188 (2012). https://doi.org/10.1016/j.bios.2012.05.033
8. Z. Luo, X. Ma, D. Yang, L. Yuwen, X. Zhu, L. Weng, L. Wang, Synthesis of highly dispersed titanium dioxide nanoclusters on reduced graphene oxide for increased glucose sensing. Carbon N. Y. **57**, 470–476 (2013). https://doi.org/10.1016/j.carbon.2013.02.020
9. N. Haghighi, R. Hallaj, A. Salimi, Immobilization of glucose oxidase onto a novel platform based on modified TiO_2 and graphene oxide, direct electrochemistry, catalytic and photocatalytic activity. Mater. Sci. Eng., C **73**, 417–424 (2017). https://doi.org/10.1016/j.msec.2016.12.015
10. S. Boobphahom, P. Rattanawaleedirojn, Y. Boonyongmaneerat, TiO_2 sol/graphene modified 3D porous Ni foam: a novel platform for enzymatic electrochemical biosensor. J. Electroanal. Chem. **833**, 133–142 (2019). https://doi.org/10.1016/j.jelechem.2018.11.031
11. S. Komathi, N. Muthuchamy, K. Lee, A. Gopalan, Fabrication of a novel dual mode cholesterol biosensor using titanium dioxide nanowire bridged 3D graphene nanostacks. Biosens. Bioelectron. **84**, 64–71 (2016). https://doi.org/10.1016/j.bios.2015.11.042

12. A.K.M. Kafi, G. Wu, A. Chen, A novel hydrogen peroxide biosensor based on the immobilization of horseradish peroxidase onto Au-modified titanium dioxide nanotube arrays. Biosens. Bioelectron. **24**, 566–571 (2008). https://doi.org/10.1016/j.bios.2008.06.004
13. J. Li, T. Han, N. Wei, J. Du, X. Zhao, Three-dimensionally ordered macroporous (3DOM) gold-nanoparticle-doped titanium dioxide (GTD) photonic crystals modified electrodes for hydrogen peroxide biosensor. Biosens. Bioelectron. **25**, 773–777 (2009). https://doi.org/10.1016/j.bios.2009.08.026
14. X. Wen, M. Long, A. Tang, Flake-like Cu_2O on TiO_2 nanotubes array as an efficient nonenzymatic H_2O_2 biosensor. J. Electroanal. Chem. **785**, 33–39 (2017). https://doi.org/10.1016/j.jelechem.2016.12.018
15. Y. Li, X. Liu, H. Yuan, D. Xiao, Glucose biosensor based on the room-temperature phosphorescence of TiO_2/SiO_2 nanocomposite. Biosens. Bioelectron. **24**, 3706–3710 (2009). https://doi.org/10.1016/j.bios.2009.05.033
16. Y. Li, P. Hsu, S. Chen, Multi-functionalized biosensor at WO_3—TiO_2 modified electrode for photoelectrocatalysis of norepinephrine and riboflavin. Sensors Actuators B. Chem. **174**, 427–435 (2012). https://doi.org/10.1016/j.snb.2012.06.061
17. Q. Huang, Y. Wang, L. Lei, Z. Xu, W. Zhang, Photoelectrochemical biosensor for acetylcholinesterase activity study based on metal oxide semiconductor nanocomposites. J. Electroanal. Chem. **781**, 377–382 (2016). https://doi.org/10.1016/j.jelechem.2016.07.007
18. N. Tio, B. Çak, E. Gökgöz, M. Özacar, A photoelectrochemical glucose and lactose biosensor consisting of gold nanoparticles, MnO_2 and g-C3N4 decorated TiO_2. Sensors Actuators B Chem. **282**, 282–289 (2019). https://doi.org/10.1016/j.snb.2018.11.064
19. J. Tian, Y. Li, J. Dong, M. Huang, J. Lu, Photoelectrochemical TiO_2 nanotube arrays biosensor for asulam determination based on in-situ generation of quantum dots. Biosens. Bioelectron. **110**, 1–7 (2018). https://doi.org/10.1016/j.bios.2018.03.038
20. P. Liu, X. Huo, Y. Tang, J. Xu, X. Liu, D.K.Y. Wong, TiO_2 nanosheet-g-C_3N_4 composite photoelectrochemical enzyme biosensor excitable by visible irradiation. Anal. Chim. Acta **984**, 86–95 (2017). https://doi.org/10.1016/j.aca.2017.06.043
21. B. Çak, M. Özacar, A self-powered photoelectrochemical glucose biosensor based on supercapacitor Co_3O_4-CNT hybrid on TiO_2. Biosens. Bioelectron. **119**, 34–41 (2018). https://doi.org/10.1016/j.bios.2018.07.049
22. X. Liu, R. Yan, J. Zhu, X. Huo, X. Wang, Development of a photoelectrochemical lactic dehydrogenase biosensor using multi-wall carbon nanotube—TiO_2 nanoparticle composite as coenzyme regeneration tool. Electrochim. Acta **173**, 260–267 (2015). https://doi.org/10.1016/j.electacta.2015.05.059
23. R. Atchudan, N. Muthuchamy, T. Nesakumar, J. Immanuel, An ultrasensitive photoelectrochemical biosensor for glucose based on bio-derived nitrogen-doped carbon sheets wrapped titanium dioxide nanoparticles. Biosens. Bioelectron. **126**, 160–169 (2019). https://doi.org/10.1016/j.bios.2018.10.049
24. N. Hao, R. Hua, S. Chen, Y. Zhang, Z. Zhou, J. Qian, Q. Liu, K. Wang, Multiple signal-amplification via Ag and TiO_2 decorated 3D nitrogen doped graphene hydrogel for fabricating sensitive label-free photoelectrochemical thrombin aptasensor. Biosens. Bioelectron. **101**, 14–20 (2018). https://doi.org/10.1016/j.bios.2017.09.014
25. D. Jiang, X. Du, D. Chen, L. Zhou, W. Chen, Y. Li, N. Hao, J. Qian, Q. Liu, K. Wang, One-pot hydrothermal route to fabricate nitrogen doped graphene/Ag-TiO_2: Efficient charge separation, and high-performance "on-off-on" switch system based photoelectrochemical biosensing. Biosens. Bioelectron. **83**, 149–155 (2016). https://doi.org/10.1016/j.bios.2016.04.042
26. N. Nesakumar, K. Thandavan, S. Sethuraman, U. Maheswari, J. Bosco, B. Rayappan, An electrochemical biosensor with nanointerface for lactate detection based on lactate dehydrogenase immobilized on zinc oxide nanorods. J. Colloid Interface Sci. **414**, 90–96 (2014). https://doi.org/10.1016/j.jcis.2013.09.052
27. D.P. Neveling, T.S. Van Den Heever, W.J. Perold, L.M.T. Dicks, A nanoforce ZnO nanowire-array biosensor for the detection and quantification of immunoglobulins. Sensors Actuators B. Chem. **203**, 102–110 (2014). https://doi.org/10.1016/j.snb.2014.06.076

28. N. Muthuchamy, R. Atchudan, T. Nesakumar, J. Immanuel, High-performance glucose biosensor based on green synthesized zinc oxide nanoparticle embedded nitrogen-doped carbon sheet. J. Electroanal. Chem. **816**, 195–204 (2018). https://doi.org/10.1016/j.jelechem.2018.03.059

29. H.A. Wahab, A.A. Salama, A.A. El Saeid, M. Willander, O. Nur, I.K. Battisha, Zinc oxide nanorods based glucose biosensor devices fabrication. Results Phys. **9**, 809–814 (2018). https://doi.org/10.1016/j.rinp.2018.02.077

30. X. Ren, D. Chen, X. Meng, F. Tang, X. Hou, D. Han, L. Zhang, Zinc oxide nanoparticles/glucose oxidase photoelectrochemical system for the fabrication of biosensor. J. Colloid Interface Sci. **334**, 183–187 (2009). https://doi.org/10.1016/j.jcis.2009.02.043

31. T. Dayakar, V.R.K.K. Bikshalu, V. Rajendar, S. Park, Novel synthesis and structural analysis of zinc oxide nanoparticles for the non enzymatic glucose biosensor. Mater. Sci. Eng. C. **75**, 1472–1479 (2017). https://doi.org/10.1016/j.msec.2017.02.032

32. C. Karuppiah, M. Velmurugan, S. Chen, A simple hydrothermal synthesis and fabrication of zinc oxide—copper oxide heterostructure for the sensitive determination of nonenzymatic glucose biosensor. Sens. Actuators B. Chem. **221**, 1299–1306 (2015). https://doi.org/10.1016/j.snb.2015.07.075

33. M. Mazaheri, H. Aashuri, A. Simchi, Three-dimensional hybrid graphene/nickel electrodes on zinc oxide nanorod arrays as non-enzymatic glucose biosensors. Sens. Actuators B. Chem. **251**, 462–471 (2017). https://doi.org/10.1016/j.snb.2017.05.062

34. A. Islam, D.W. Morton, B.B. Johnson, B. Mainali, M.J. Angove, Manganese oxides and their application to metal ion and contaminant removal from wastewater. J. Water Process Eng. **26**, 264–280 (2018). https://doi.org/10.1016/j.jwpe.2018.10.018

35. W. Bai, X. Zhang, S. Zhang, Q. Sheng, J. Zheng, Acidification of manganese dioxide for ultrasensitive electrochemical sensing of hydrogen peroxide in living cells. Sens. Actuators B. Chem. **242**, 718–727 (2017). https://doi.org/10.1016/j.snb.2016.11.125

36. L. Li, Z. Du, S. Liu, Q. Hao, Y. Wang, Q. Li, T. Wang, A novel nonenzymatic hydrogen peroxide sensor based on MnO_2/graphene oxide nanocomposite. Talanta **82**, 1637–1641 (2010). https://doi.org/10.1016/j.talanta.2010.07.020

37. A.O. Idris, O.A. Arotiba, Towards cancer diagnostics—an α-feto protein electrochemical immunosensor on a manganese (IV) oxide/gold nanocomposite immobilisation layer. RSC Adv. **8**, 30683–30691 (2018). https://doi.org/10.1039/C8RA06135A

38. L. Li, J. Xu, X. Zheng, C. Ma, X. Song, S. Ge, J. Yu, M. Yan, Growth of gold-manganese oxide nanostructures on a 3D origami device for glucose-oxidase label based electrochemical immunosensor. Biosens. Bioelectron. **61**, 76–82 (2014). https://doi.org/10.1016/j.bios.2014.05.012

39. M. Tu, H. Chen, Y. Wang, S.M. Moochhala, P. Alagappan, B. Liedberg, Immunosensor based on carbon nanotube/manganese dioxide electrochemical tags. Anal. Chim. Acta **853**, 228–233 (2015). https://doi.org/10.1016/j.aca.2014.09.050

40. Y. Bai, Y. Du, J. Xu, H. Chen, Choline biosensors based on a bi-electrocatalytic property of $MnO2$ nanoparticles modified electrodes to H2O2. Electrochem. Commun. **9**, 2611–2616 (2007). https://doi.org/10.1016/j.elecom.2007.08.013

41. G. Yu, Q. Zhao, W. Wu, X. Wei, Q. Lu, A facile and practical biosensor for choline based on manganese dioxide nanoparticles synthesized in-situ at the surface of electrode by one-step electrodeposition. Talanta **146**, 707–713 (2016). https://doi.org/10.1016/j.talanta.2015.06.037

42. G.L. Luque, N.F. Ferreyra, A.G. Leyva, G.A. Rivas, Characterization of carbon paste electrodes modified with manganese based perovskites-type oxides from the amperometric determination of hydrogen peroxide. Sens. Actuators B Chem. **142**, 331–336 (2009). https://doi.org/10.1016/j.snb.2009.07.038

43. C. Zhang, Y. Zhang, Z. Miao, M. Ma, X. Du, J. Lin, B. Han, S. Takahashi, J.I. Anzai, Q. Chen, Dual-function amperometric sensors based on poly(diallydimethylammoniun chloride)-functionalized reduced graphene oxide/manganese dioxide/gold nanoparticles nanocomposite. Sens. Actuators, B Chem. **222**, 663–673 (2016). https://doi.org/10.1016/j.snb.2015.08.114

44. X. Yang, X. Chen, X. Zhang, W. Yang, D.G. Evans, Intercalation of methylene blue into layered manganese oxide and application of the resulting material in a reagentless hydrogen peroxide biosensor. Sens. Actuators B. **129**, 784–789 (2008). https://doi.org/10.1016/j.snb.2007.09.063

45. P. Zhang, D. Guo, Q. Li, Manganese oxide ultrathin nanosheets sensors for non-enzymatic detection of H_2O_2. Mater. Lett. **125**, 202–205 (2014). https://doi.org/10.1016/j.matlet.2014.03.172

46. E. Turkusic, J. Kalcher, E. Kahrovic, N.W. Beyene, Amperometric determination of bonded glucose with an MnO_2 and glucose oxidase bulk-modified screen-printed electrode using flow-injection analysis. Talanta **65**, 559–564 (2005). https://doi.org/10.1016/j.talanta.2004.07.023

47. V. Vukojevi, S. Djurdji, M. Fabián, A. Samphao, K. Kalcher, D.M. Stankovi, Enzymatic glucose biosensor based on manganese dioxide nanoparticles decorated on graphene nanoribbons. J. Electroanal. Chem. **823**, 610–616 (2018). https://doi.org/10.1016/j.jelechem.2018.07.013

48. X. Cui, G. Liu, Y. Lin, Amperometric biosensors based on carbon paste electrodes modified with nanostructured mixed-valence manganese oxides and glucose oxidase. Nanomed. Nanotechnol. Biol. Med. **1**, 130–135 (2005). https://doi.org/10.1016/j.nano.2005.03.005

49. L. Zhang, S. Yuan, L. Yang, An enzymatic glucose biosensor based on a glassy carbon electrode modified with manganese dioxide nanowires. Microchim. Acta **180**, 627–633 (2013). https://doi.org/10.1007/s00604-013-0968-9

50. A.P. Mahajan, S.B. Kondawar, R.P. Mahore, B.H. Meshram, Polyaniline/MnO_2 nanocomposites based stainless steel electrode modified enzymatic urease biosensor. Procedia Mater. Sci. **10**, 699–705 (2015). https://doi.org/10.1016/j.mspro.2015.06.075

51. M. Hasanzadeh, N. Shadjou, M. De, Iron and iron-oxide magnetic nanoparticles as signal-amplification elements in electrochemical biosensing. Trends Anal. Chem. **72**, 1–9 (2015). https://doi.org/10.1016/j.trac.2015.03.016

52. S. Siddique, U. Rahman, M. Tauseef, K. Sultana, W. Rehman, M. Yaqoob, M. Hassan, M. Farooq, N. Sultana, Single step growth of iron oxide nanoparticles and their use as glucose biosensor. Results Phys. **7**, 4451–4456 (2017). https://doi.org/10.1016/j.rinp.2017.11.001

53. S. Kumar, M. Umar, A. Sai, S. Kumar, S. Augustine, S. Srivastava, B.D. Malhotra, Electrochemical paper based cancer biosensor using iron oxide nanoparticles decorated PEDOT: PSS. Anal. Chim. Acta **1056**, 135–145 (2019). https://doi.org/10.1016/j.aca.2018.12.053

54. H. Zhang, Y. Sun, S. Gao, H. Zhang, J. Zhang, Y. Bai, D. Song, Studies of gold nanorod-iron oxide nanohybrids for immunoassay based on SPR biosensor. Talanta **125**, 29–35 (2014). https://doi.org/10.1016/j.talanta.2014.02.036

55. N. Mohamad, K. Abdul, Z. Lockman, Physical and electrochemical properties of iron oxide nanoparticles-modified electrode for amperometric glucose detection. Electrochim. Acta **248**, 160–168 (2017). https://doi.org/10.1016/j.electacta.2017.07.097

56. N. Mohamad, Z. Lockman, K. Abdul, Study of ITO glass electrode modified with iron oxide nanoparticles and Nafion for glucose biosensor application. Procedia Chem. **19**, 50–56 (2016). https://doi.org/10.1016/j.proche.2016.03.116

57. A. Kaushik, R. Khan, P.R. Solanki, P. Pandey, J. Alam, S. Ahmad, B.D. Malhotra, Iron oxide nanoparticles—chitosan composite based glucose biosensor. Biosens. Bioelectron. **24**, 676–683 (2008). https://doi.org/10.1016/j.bios.2008.06.032

58. Y. Wang, X. Liu, X. Xu, Y. Yang, L. Huang, Z. He, Y. Xu, Preparation and characterization of reduced graphene oxide/Fe_3O_4 nanocomposite by a facile in-situ deposition method for glucose biosensor applications. Mater. Res. Bull. **101**, 340–346 (2018). https://doi.org/10.1016/j.materresbull.2018.01.035

59. S. Pakapongpan, R.P. Poo-arporn, Self-assembly of glucose oxidase on reduced graphene oxide-magnetic nanoparticles nanocomposite-based direct electrochemistry for reagentless glucose biosensor. Mater. Sci. Eng., C **76**, 398–405 (2017). https://doi.org/10.1016/j.msec.2017.03.031

60. J. Li, R. Yuan, Y. Chai, X. Che, Fabrication of a novel glucose biosensor based on Pt nanoparticles-decorated iron oxide-multiwall carbon nanotubes magnetic composite. J. Mol. Catal. B Enzym. **66**, 8–14 (2010). https://doi.org/10.1016/j.molcatb.2010.03.005

61. A. Kaushik, P.R. Solanki, A.A. Ansari, G. Sumana, S. Ahmad, B.D. Malhotra, Iron oxide-chitosan nanobiocomposite for urea sensor. Sens. Actuators B Chem. **138**, 572–580 (2009). https://doi.org/10.1016/j.snb.2009.02.005

62. E. Çevik, S. Mehmet, Potentiometric urea biosensor based on poly (glycidylmethacrylate)-grafted iron oxide nanoparticles. Curr. Appl. Phys. **13**, 280–286 (2013). https://doi.org/10.1016/j.cap.2012.07.025

Chapter 7
Metal Oxide Nanomaterials for Electrochemical Detection of Heavy Metals in Water

Seyi Philemon Akanji, Onoyivwe Monday Ama, Suprakas Sinha Ray, and Peter Ogbemudia Osifo

Abstract Metal oxide nanomaterials (MONMs) have gained much research attention as prospective materials of preference for modifying electrode surfaces intended for electrochemical sensing of heavy metal ions. MONMs have attracted much research interest because of their unique characteristics at nano dimensions compared to bulk species. In general, the unique electronic properties, high electron transfer kinetics, semi-conducting properties among others make them exceptionally smart materials for electrochemical reaction. Moreover, early detection of toxic metal ions in our water bodies will help to improve the standard of human living in the environment. Owing to this effect, the present chapter deals with the utilization of electrochemical techniques such as voltammetry and potentiometry among other electrochemical techniques for sensing of heavy metal ions in water bodies. The chapter also address the progress in the application of different metal oxide nanomaterial in terms of their frameworks, adsorption capacities, surface structure, chemical functional group attached, and how their use as preference of electrode materials affect electrochemical sensing of heavy metal ions in water.

7.1 Introduction

The importance of heavy metals such as copper, lead, iron, zinc among others, cannot be underestimated because of the vital role they play in the consumption of portable electronic devices [1] and in living entities—being essential in small amounts for

S. P. Akanji (✉) · O. M. Ama · S. S. Ray
Department of Chemical Science, University of Johannesburg, Doornfontein, 2028, Johannesburg, South Africa
e-mail: philemonakanji73@gmail.com

O. M. Ama · S. S. Ray
DST-CSIR National Center for Nanostructured Materials, Council for Scientific and Industrial Research, Pretoria 0001, South Africa

O. M. Ama · P. O. Osifo
Department of Chemical Engineering, Vaal University of Technology, Private Mail Bag X021, Vanderbijlpark 1900, South Africa

© Springer Nature Switzerland AG 2020
O. M. Ama and S. S. Ray (eds.), *Nanostructured Metal-Oxide Electrode Materials for Water Purification*, Engineering Materials,
https://doi.org/10.1007/978-3-030-43346-8_7

maintenance of metabolic activities [2]. As an example, the production of modern electronic devices such as smartphones and tablets requires the use of over 60 different metals ranging from silver, gold, platinum and copper to rare earth elements [1, 3]. Also, manganese is a carrier of oxygen in the tissue redox processes [4]. Nevertheless, excess or slight amount of very toxic metals (i.e. copper, mercury etc.) can lead to severe health and environmental complications. The ingestion of such metals into the body system can damage the central nervous system, kidneys, liver, skin, bones and teeth [2]. For example, the ingestion of fruit improperly sprayed with copper insecticide can result into transitional poisoning [4].

Major sources of heavy metals include the dependency of metals in the consumption of portable electronic devices [1], agriculture, and mining industry [2]. The release of heavy metals into water bodies and terrestrial systems may be caused by such activities (mining, urbanisation) which are further transferred to plants, animals and humans. As an example, it was reported that over 200 people died of mercury poisoning—called Minamata disease—in Japan in the gulf of Minamata, after eating fish contaminated with mercury compounds [4]. Hence, it is imperative to develop sensors which can detect heavy metals early enough both in living structures and the environment at large [2].

Spectroscopic techniques such as atomic absorption spectrometry (AAS), inductively coupled plasma optical mass spectrometry (ICP-OES) [2, 5], inductively coupled plasma mass spectroscopy (ICP-MS) [6] are normally being used for analysis of heavy metal ions. These spectroscopic methods are multi-tasking in the sense that analysis of multiple metal ions can be produce together within a short period. Also, they can detect elements as very minute as femtomolar amounts [6]. The possible shortcomings of using these spectroscopic methods is found in their complexities, high-priced and highly skilled workforces required to carry out the operational procedures [2, 6]. The use of optical techniques such as spectrophotometric measurements have also been reported for the detection of heavy metal ions [2, 6]. Therefore, the optical techniques, once more, involve complex and costly equipment coupled with lasers, etc. that need great power operations and precision, which are not suitable for in-field approach. These reasons led to continuing exploration on the improvement of fast, economical and convenient methods for investigation of heavy metal ions [2, 6, 7].

Electrochemistry methods when compared to spectroscopic and optical methods offer numerous benefits such as price, easiness, user-friendly and the prospect of in-field operation [2, 8]. The most described methods among other electrochemical methods for detection of metal ions are voltammetry and potentiometry [2]. In voltammetry, the original current at nil state of an electrochemical cell is perturbed at constant time or varying potential at the surface of an electrode and quantifying the subsequent current. Generally, stripping methods (such as anodic stripping voltammetry, ASV) amongst the diverse kinds of voltammetric procedures, are frequently utilized for trace determination of heavy metals owing to high sensitivity, selectivity, pre-concentration as well as determination steps in one single process [2, 6, 9]. One of the characteristic feature of stripping techniques is their ability to combine with linear scans or modulations (differential pulse ASV (DPASV)) and square wave ASV

(SWASV)). The significance of using ASV with linear scans or modulations during determination of heavy metal ions is a low detection limit with a high sensitivity ($\sim 10^{-6}$ to 10^{-11} M) when compared to linear sweep voltammetry (LSV) [10]. In potentiometry, there is a change in potential at zero current, which is an indication of information about the sample composition. This change will be observed at two electrodes—ion selective electrode (ISE) and field-effect transistor (FET). ISEs and FETs are the two main devices that fall within the class of potentiometry [2, 6]. In addition to the merits of electrochemical methods, they have lower sensitivity and Limits of detection (LOD) in comparison to other spectroscopic and optical techniques, although, further improvements in the application of electrochemical techniques is also required to improve their performance towards the detection of heavy metal ions [6].

The developments in the use of electrochemical techniques demand the selection of proper electrode materials for investigation of heavy metal ions. The structure of a surface and the functional groups of the attached chemical material can also affect the performance of an electrochemical sensor aside the attributes possessed by electrode materials [11]. In recent times, attributes such as high electron transfer kinetics, large surface area [2, 11] unique electronic properties, chemical and mechanical properties [2] have made the use of nanomaterials as suitable electrode material. These properties have also made them exceptionally smart materials for electrochemical reaction, hence gaining remarkable attention. To date, several nanomaterials have been explored in the preparation of electrochemical sensors for metal ions investigation and these comprises gold nanoparticles [12–15], carbon nanotubes [11, 16–18], graphene (oxide) composites [5, 19, 20] etc. Aside these conventional nanostructured materials, several metal oxides [10, 11] (iron oxide, mesoporous nickel oxide nanosheets, titanium oxide, porous magnesium oxide nanoflower, zirconium oxide [21], manganese oxide, cobalt oxide [22] etc. have been reported to be prospective materials of possibility for modifying the surface of electrodes for use in the design of electrochemical sensors. Enhanced electrochemical performance can be achieved when metal oxide nanomaterials are mixed together with other materials to form nanocomposites [11]. However, there is a need to adopt a simple approach when combining metal oxide nanomaterial with other material such as the use of simple surface modification to check agglomeration [11].

This present chapter will focus on some published work on the efficient and effective use of some metal oxide nanomaterials for electrochemical investigation of heavy metal ions in water.

7.2 Kinds of Metal Oxide Nanomaterials for Electrode Modification and Their Electrochemical Applications

Various metal oxide-based nanomaterials have been reported for the electrodes modification for sensing purposes [22, 23]. Metal oxide-based nanomaterials have diverse frameworks namely spherical, nanowires, nanotubes, nanorods, etc. Metal oxide-based nanomaterials also have excellent properties namely mechanical, magnetic, catalytic, optical, and so on. Hence, they play significant role in designing electrochemical sensors for analysis of metal ions in water.

7.2.1 TiO$_2$ Nanomaterial-Modified Electrodes

TiO$_2$ nanoparticles were incorporated into multi-walled carbon nanotubes and a different cationic surfactant, 3-dehydroabietylamine-2-hydroxypropyl trimethylammonium chloride (DHAHPTMA) [24]. Nanocomposites are used to remodel a glassy carbon electrode, TiO$_2$/MWCNTS/DHAHPTMA/GCE. The TiO$_2$/MWCNTS/DHAHPTMA/GCE was exploited to electrochemically detect trace Hg^{2+} by means of linear sweep anodic stripping voltammetry (LSASV). The composite electrode displayed a great selectivity for Hg^{2+} in the existence of frequent prospective interfering metal ions (Pb^{2+}, Cu^{2+}, Cd^{2+}, Zn^{2+}, Ni^{2+}, Mn^{2+}) due to a high adsorption of the composites (TiO$_2$/MWCNTS/DHAHPTMA) complex film. The LSAS peak current with the concentration range of Hg^{2+} reported between 0.1 and 100 μmol L^{-1} gave a detection limit of 0.025 μmol L^{-1}. This sensor was also applied for Hg^{2+} determination in stream water and industrial left-over water samples.

Ramezani and co-researchers created an electrochemical sensor for investigating Cd^{2+} constructed on an improved TiO$_2$ NPs carbon paste electrode (CPE) refined via 1,2-bis-[o-aminophenyl thio] ethane (APTE) as an innovative receptor [25]. The TiO$_2$ NPs-APTE-CPE was formulated by intermixing graphite powder with a preferred quantity of chemically modified APTE ligand and TiO$_2$ NPs. The whole blend was uniformly prepared with an ideal quantity of acetone, allowed to vapourize spare solvent at ambient condition was filled in a polyethylene tube (2.5 mm diameter). The sensor demonstrated a linear range of 2.9 nM–4.6 μM as well as a detection limit (S/N = 3) of 2.0 nM after 10 min preconcentration. The results obtained showed that APTE synthetic compound is a promising material for modification of electrochemical sensors due to its high affinity toward Cd^{2+}. Thus, APTE is a suitable CPE modifier which produces a synergistic effect when combined with TiO$_2$ NPs for Cd^{2+} sensing as compared to the performance of other sensors reported for Cd^{2+} sensing in the electrochemical literature [25].

An electrochemical sensor comprising graphene, nitrogen-doped and gold nanoparticles functionalized with 2, 2′–((1E)–((4-((2-mercaptoethyl) thio)–1, 2–phenylene) bis (azanylylidene)) bis (methanylylidene)) diphenol (ETBD) and Fe$_3$O$_4$–TiO$_2$ core-shell nanoparticles on a GCE as a platform for detecting Pb(II) was described by Liu and co-workers [26]. Fe$_3$O$_4$–TiO$_2$ is utilized because it supports chemical firmness, electrical communication and great adsorption ability for Pb^{2+}. Nitrogen-doped graphene (NG) was employed as the support material, which speeds up the rate of electron shift for signal improvement as well as creates spots for Fe$_3$O$_4$–TiO$_2$ and gold nanoparticles (Au NPs). Au NPs were initiated into NG to boost the shift of electron as well as operate as a channel linking the inorganic and organic compounds (ETBD) through stable S-Au bond, bringing about loading additional quantities of ETBD in addition to the creation of extra firm and clear conductive positions. Electrochemical response of the proposed sensor, Fe_2O_3–TiO_2–NG–Au NPs–ETBD/GCE, toward Pb(II) detection was investigated using square wave voltammetry (SWV). The LOD and quantification for Pb^{2+} of the proposed sensor were calculated to be 7.5×10^{-13} and 2.5×10^{-12} mol/L over a varied direct range from 4×10^{-13} mol/L to 2×10^{-8} mol/L. A great extent of the organic compounds, ETBD, led to a direct conjugation of a great quantity of Pb^{2+} with ETBD to create [Pb(II)-ETBD]$^{n+}$ aimed at label-free recognition making Fe_2O_3–TiO_2–NG–AuNPs–ETBD demonstrates great sensitivity and selectivity towards Pb(II) [26].

7.2.2 Fe_2O_3, Fe_2O_4 and Fe_3O_4 Nanomaterial-Modified Electrodes

The adsorption capacity of the surface of an electrode performs a vital part in the electrochemical sensing of heavy metals [22, 27]. The high adsorption capacity and super-paramagnetism properties of Ferrite (MFe_2O_4; M = Fe, Mn, Zn, or Co) nanomaterials is well documented [22]. Li et al. [28] reported the performance of two different morphologies of Fe_2O_3—nanorods and hollow nanocubes—towards the electroanalysis of Pb(II). Their report showed that the surface area and the exposed sites of nanomaterials perform a vital function in the adsorption of metal ions towards various electrochemical detection response. Thus, Fe_2O_3 with two different morphologies were synthesized as reported [28] and used to modify bare GCE electrodes towards the electroanalysis of Pb(II). The LOD for Pb(II) on Fe_2O_3 nanorods ($0.0034\ \mu M$) was found to be far lesser compared to hollow nanocubes ($0.083\ \mu M$), indicating the exceptional sensing characteristic of Fe_2O_3 nanorods. Besides, Fe_2O_3 nanorods are found to be more sensitive ($109.67\ \mu A\ \mu M^{-1}$) compared to hollow nanocubes ($17.68\ \mu A\ \mu M^{-1}$). Thus it was proposed that an enrichment of the active spot on the surface of nanorods may perhaps increase the adsorption of all heavy metal ions, further reinforcing the electrochemical signal [28].

Lee and co-workers described a Fe_2O_3/graphene/bismuth (Fe_2O_3/G/Bi) nanocomposite modified GCE [19] for the concurrent detection of Zn^{2+}, Cd^{2+} and Pb^{2+}. A solvent-free heating degradation scheme was employed in the preparation of the nanocomposite electrode materials as reported [19]. The proposed sensor demonstrated a direct range of 1–100 $\mu g\ L^{-1}$ for Zn^{2+}, Cd^{2+} and Pb^{2+} as well as LODs of 0.11 $\mu g\ L^{-1}$, 0.08 $\mu g\ L^{-1}$ and 0.07 $\mu g\ L^{-1}$ (S/N = 3). The Fe_2O_3/G/Bi composite electrode is effectively employed in trace determination of metal ions in actual samples.

A graphene oxide integrated mesoporous $MnFe_2O_4$ nanocomposites electrode ($MnFe_2O_4$/GO) was also reported by Zhou et al. for sensing of Pb^{2+} [5]. The electrochemical responses of the fabricated electrode ($MnFe_2O_4$/GO) were investigated by means of square wave anodic stripping voltammetry (SWASV). The LOD (S/N = 3) of Pb^{2+} was $0.0883\ \mu M$. To further check the recognition susceptibility of the sensor ($MnFe_2O_4$/GO), further experiments were carried out to investigate its sensitivity towards Cu^{2+}, Cd^{2+} and Hg^{2+}. Measurements by means of SWASV technique showed their LODs to be 0.778 μM Cu(II), 0.0997 μM Cd(II) and 1.16 μM Hg(II). By comparing the three-analyte ions with Pb^{2+}, Pb^{2+} has the lowest LOD and highest sensitivity; an indication that $MnFe_2O_4$/GO nanocomposites modified GCE demonstrated enhanced electrochemical activity to Pb^{2+} in comparison to other three heavy metal ions. This was reported to be because as a result of the cooperative influence of GO and mesoporous $MnFe_2O_4$ on the deposition of Pb^{2+} [5].

Xiong and co-researchers devise a reduced graphene oxide—Fe_3O_4 nanocomposites (rGO-Fe_3O_4) electrode for investigating Cd^{2+}, Pb^{2+} and Hg^{2+} [29]. The rGO-Fe_3O_4 nanocomposites were developed by means of one-pot synthesis method [29]. The nanocomposites were employed for the modification of GCE for detection of Cd^{2+}, Pb^{2+} and Hg^{2+} separately and concurrently. During separate analysis of metal ions, the LOD was 8 nM in the range of 0.3–3 μM Cd(II). For Pb^{2+}, the LOD was 6 nM in the range of 0.2–1.3 μM whereas for Hg^{2+}, the LOD was 4 nM in the range of 0.4–1.8 μM. Hg(II) has the lowest LOD amongst the three analyte ions which was attributed to improved similarity interaction between Hg(II) and rGO-Fe_3O_4 nanocomposites [29]. On the other hand, SWASV feedbacks of the rGO-Fe_3O_4 electrode for the concurrent detection of Cd^{2+}, Pb^{2+} and Hg^{2+} in the range 0.1–1.7 μM gave LODs of 28, 8 and 17 nM for Cd^{2+}, Pb^{2+} and Hg^{2+}, correspondingly. By observation, the sensitivities of the sensor for Cd(II), Pb(II) and Hg(II) were lower than those of corresponding separate detection. There was also a rise in the LODs for Cd^{2+}, Hg^{2+} and Pb^{2+} compared to those of separate detection. Therefore, the anticipated sensor was effectively utilized for deduction of the three analytes of interest separately and concurrently [29].

The outstanding features of magnetite/reduced graphene oxide nanomaterials such as large surface area, remarkable conductibility, durability, powerful magnetic and broad electrochemical window was exploited by Sun and co-researchers to develop Fe_3O_4/rGO nanocomposite. The nanocomposite (Fe_3O_4/rGO) which were of different shapes (spherical, rod and band) was used to develop a Fe_3O_4/rGO-modified GCE [30]. The different nanostructures of the designed sensor were used to detect Pb^{2+} using SWASV. Prevalent metal ions namely Cd^{2+} and Cu^{2+} were also examined to measure the selectivity of the band Fe_3O_4/rGO-modified GCE. Fe_3O_4/rGO nanocomposite was synthesized from one-step solution co-precipitation method. The framework (spherical, rod and band) of Fe_3O_4 in the nanocomposites were cautiously regulated by changing the ratio of Fe^{2+} to Fe^{3+} moles in the presence of rGO modified polyvinylpyrrolidone (PVP). The devised spherical Fe_3O_4/rGO-modified GCE demonstrated a sensitivity of ca. 7.4 μA/μM as well as a LOD of 0.073 μM (3σ method) in the range 0.7–1.2 μM towards Pb^{2+} detection. Rod Fe_3O_4/rGO-modified GCE demonstrated a sensitivity of ca. 2.4 μA/μM as well as a LOD of 0.033 μM (3σ method) in the range 0.8–1.2 μM. Band Fe_3O_4/rGO-modified GCE demonstrated a sensitivity of ca.13.6 μA/μM as well as a LOD of 0.17 μM (3σ method) in the range 0.4–1.5 μM towards Pb^{2+} detection. From observation, the preferred sensitivity was evidence on band Fe_3O_4/rGO-modified GCE toward Pb^{2+} detection amongst the three devised nanomaterial modified electrodes with no significant differences in their LODs, the selectivity of the band Fe_3O_4/rGO-modified GCE was investigated towards other metal ions—Cd^{2+} and Cu^{2+} as reported [30]. A LOD of 0.04 μM (3σ method) as well as a sensitivity of 4.35 μA/μM recorded for SWASV responses of the band Fe_3O_4/rGO-modified GCE in respect to Cd^{2+} sensing in the range 0.4–1.1 μM. On the other hand, band Fe_3O_4/rGO-modified GCE demonstrated a LOD of 0.05 μM (3σ method) as well as a sensitivity of 10.1 μA/μM in the range 0.5–1.5 μM towards Cu^{2+} sensing. The selectivity behaviour of Fe_3O_4/rGO-modified GCE in respect to the Pb^{2+}, Cu^{2+} and Cd^{2+} detection could be summarised as: Pb^{2+} > Cu^{2+} > Cd^{2+}. Conclusively, the results and performance of the various modified glassy carbon electrodes has actually showed how framework of electrode materials affect the sensing performance in respect to metal ions detection [30].

7.2.3 SnO₂ Nanomaterial-Modified Electrodes

A simple hydrothermal process was employed in the preparation of SnO_2/QDs—which also serve as a medium of electron transfer—in the preparation of gold electrode (SnO_2/Nafion/Au) in the presence of nafion solution for Cd^{2+} sensing [31]. The nature of the SnO_2/Nafion/Au electrode was explored by carrying out series of cyclic voltammetry (CV) measurements at a scan rate of 100 mV/s in 5–45 ppm Cd^{2+} in citrate buffer solution (0.1 M, pH = 5). Chrono-amperometry measurements were employed in the selectivity and interference studies. The anticipated sensor displayed a sensitivity of ~77.5 × 10^2 nA ppm^{-1} cm^{-2} as well as a LOD of ~0.5 ppm. The observed response time was <2 s [31].

Yang and co-workers utilized the attributes of green solvent room temperature ionic liquids (RTILS)—such as conductivity, broad electrochemical windows, tunableness etc.—to devise an amino—based SnO_2 nanowire bundles—RTILS nanocomposite GCE. The resulting modified NH_2/SnO_2-RTIL/GCE was utilized in ultra-trace detection of Cd^{2+} by means of SWASV [32]. The sensor displayed a LOD of 0.0054 μM (3σ method) as well as a sensitivity of 124.03 ± 3.75 μA μM^{-1}. The sensor was also utilized in effective determination of Cd^{2+} in actual water sample [32].

Cui and co-researchers described a 2-amino benzothiazole and 2-amino-4-thiazoleacetic acid derivative graphene enhanced with fluorine, chlorine and iodine on SnO_2 nanoparticles applied on a bare GCE to achieve F–SnO_2/T/RGO/GCE, Cl–SnO_2/T/RGO/GCE and Cl–SnO_2/T/RGO/GCE for sensing Cu^{2+}, Cd^{2+}, Cu^{2+} and Hg^{2+} [33]. Each of the derivative sensors was used for sensing Cu^{2+} as well as concurrent detection of Cd^{2+}, Cu^{2+} and Hg^{2+}. Cyclic voltammetry measurements showed F–SnO_2/T/RGO electrode as the best amongst the three electrodes owing to the fact that it can adsorb and transport more Cu^{2+} in the range 0–1000 nM; hence, F–SnO_2/T/RGO electrode was utilize for latter experiments. Result of the differential pulse voltammetry (DPV) measurement revealed the LOD of Cu^{2+} to be 0.3 nM (S/N = 3) in the range 2–1000 nM at F–SnO_2/T/RGO/GCE. The statistical equation was I

$(\mu A) = 0.014\, C_{Cu2+} + 0.827$. Additional use of F–SnO$_2$/T/RGO/GCE in the detection of Cd^{2+}, Cu^{2+} and Hg^{2+} concurrently at different redox potentials led to LOD of 5 nM, 3 nM and 5 nM (S/N = 3) in the range 0–2000 nM. Therefore, the requirements for measurement of Cu^{2+} was different from that of concurrent measurements of Cd^{2+} and Hg^{2+} statistical equation of I (μA) = 0.009 C$_{Cd2+}$ + 0.409, I (μA) = 0.034 C$_{Cu2+}$ − 0.827 and I (μA) = 0.014 C$_{Hg2+}$ − 0.636 were both derived for the two different measurements, proving that F–SnO$_2$/T/RGO/GCE could be utilized for concurrent measurement of Cd^{2+}, Cu^{2+} and Hg^{2+} [33].

7.2.4 ZnO Nanomaterial-Modified Electrodes

Yuan-Yuan and co-workers explored the semiconducting property as well as the catalytic ability of ZnO via intermixing with reduced graphene oxide (RGO) to obtain a ZnO/RGO nanocomposite for the analysis of Pb^{2+} [34]. The ZnO/RGO nanocomposites were carefully prepared via electrospinning and thermal decomposition of Zn(Ac)$_2$-polyacrylonitrile-polyvinyl pyrrolidone (Zn(Ac)$_2$PAN/PVP) precursor to obtain ZnO nanotubes and then mixed with RGO to devise the anticipated sensor, Zn-RGO/GCE. SWSV response of the Zn-RGO/GCE gave a detection limit of 4.8 × 10^{-10} M (S/N ≥ 3) in the concentration range of 2.4 × 10^{-9}–4.8 × 10^{-7} M. The anticipated sensor was also utilized for Pb^{2+} in actual water sample [34].

Yukird and co-researchers described a sensor made of ZnO and graphene (G) nanocomposite for concurrent analysis of Cd^{2+} and Pb^{2+} [35]. Thermal pyrolysis procedure was utilized for the synthesis of ZnO nanorod and was mixed with graphene solution through colloidal coagulation influence to devise a screen-printed carbon electrode (SPCE) as the anticipated sensor (ZnO–G/SPCE). Electrochemical activity of the ZnO–G/SPCE was first controlled by exploiting the mixing ratio of ZnO and graphene varied at 90:10, 80:20, 70:30, 60:40, and 50:50 respectively. Overall, a dimension of 80:20 was selected for latter experiment owing to the fact that the current response of the anodic stripping voltammetry (ASV) detection of 50 μg L^{-1} Cd^{2+} and Pb^{2+} (Fig. 7.1a right) increases with increasing dimension of graphene between 10 and 20%. This confirms the positive influence of graphene on the activity of the system. On the other hand, more than 20% of the graphene content resulted in a decrease in the current response owing to re-accumulation of graphene nanosheet demonstrated by the high background current response. Moreover, ASV studies of ZnO–G concentration in the range 1–4 mg/mL (Fig. 7.1b, Left) resulted in a pre-selected concentration of 2 mg/mL for both analytes owing to an increase in the current response (Fig. 7.1b, Right) with increasing concentration of ZnO–G. This connotes the positive influence of ZnO–G on the sensitivity of the system.

In addition, concurrent determination of Cd^{2+} and Pb^{2+} on the ZnO–G sensor by means of ASV in the range 10–200 μg L^{-1} (Fig. 7.2a) gave R^2 values of 0.9968 (Cd^{2+}) and 0.9986 (Pb^{2+}) respectively (Fig. 7.2b, c). Their respective calculated LODs were 0.6 μg L^{-1} (Cd^{2+}) and 0.8 μg L^{-1} (Pb^{2+}) (LOD = 3SD$_b$/M, where SD$_b$

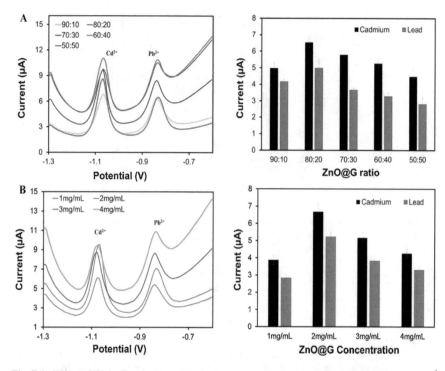

Fig. 7.1 Effect of ZnO–G ratio (**a**) and ZnO–G concentration (**b**) on stripping peak of 50 μg L^{-1} Cd^{2+} and Pb^{2+} in 0.1 M acetate buffer solution (pH 4.5). The error bars correspond to standard deviation obtained from 3 measurements (n = 3). Adapted from Yukird et al. [35] with permission

is the standard deviation from the blank signal, and M is slope of calibration curve of the standard) [35].

7.2.5 MnO, MnO$_2$ Nanomaterial-Modified Electrodes

The activity of a ternary nanocomposite comprising chitosan (Chit), multiwalled carbon nanotubes (MWCNTS) and manganese oxide (Chit/MWCNTs/MnOx) toward chromium (III) detection was studied by Salimi and co-researchers [36]. MnOx nanoflakes were deposited on the surface of a bare GCE uniformly modified with film of Chit/MWCNTs (GC/Chit/MWCNTs/MnOxNP). Cyclic voltammetry and amperometry response of the GC/Chit/MWCNTs/MnOxNP sensor resulted in a LOD of 0.3 μM in the range 3–200 μM as well as a sensitivity of 18.7 nA μM^{-1} [36].

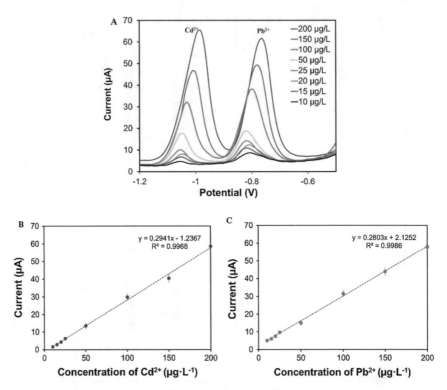

Fig. 7.2 Anodic stripping voltammograms of Cd^{2+} and Pb^{2+} in the range 10-200 $\mu g\ L^{-1}$ (**a**) the calibration chart of Cd^{2+} concentration against the current response (**b**) and the calibration chart of Pb^{2+} concentration against the current response (**c**). The error bars conform to the standard deviation taken from 3 quantifications (n = 3). Adapted from Yukird et al. [35] with permission

Fayazi and co-researchers described a recent magnetic electrochemical sensor made of halloysite nanotubes–iron oxide–manganese oxide nanocomposite (HNTs–Fe_3O_4–MnO_2), an extractant for mercury(II) ions detection [37]. HNT is an alumi-nosilicate type of clay, $Al_2(OH)_4Si_2O_5 \cdot 2H_2O$, whose chemical composition is sim-ilar to the polytype of kaolinite clays—kaolinite, dickite or nacrite—although the framework of the HNT crystals differs from the kaolinite polytypes [37, 38]. Simple chemical precipitation as well as hydrothermal method were adopted in the synthe-sis of HNTs–Fe_3O_4–MnO_2 [37, 39]. Figure 7.3 shows the representative design of the integrated process in the synthesis of HNTs–Fe_3O_4–MnO_2. The HNTs–Fe_3O_4–MnO_2 nanocomposites were employed in the preparation of a magnetic carbon paste electrode (MCPE), the anticipated sensor (HNTs–Fe_3O_4–MnO_2/MCPE) for the anal-ysis of Hg^{2+}. Electrochemical response by means of differential pulse voltammetry (DPV) displayed a LOD of 0.2 $\mu g\ L^{-1}$ of Hg^{2+} in the range 0.5–150 $\mu g\ L^{-1}$. The presence of MnO_2 on the magnetic HNTs in the nanocomposite mixture intensely improved the extraction and detection of Hg^{2+}. The anticipated sensor was likewise utilized for estimation of Hg^{2+} in actual water samples [37].

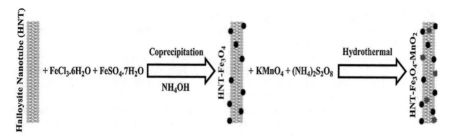

Fig. 7.3 A representative design of the integrated process in the synthesis of HNTs–Fe_3O_4–MnO_2. Adapted from Fayazi et al. [37] with permission

7.3 Deductions and Forthcoming Outlook

This topic obviously identified the choices of metal oxide nanomaterial, how the features of the electrode materials such as magnetic properties, surface structure and attached chemical functional groups can affect the performance of electrochemical sensors towards the sensing of heavy metal ions. These classes of materials despite their uniqueness regarding their structures and attributes are implemented in wide range of applications. The topic apparently found that the features of metal oxide nanomaterials are fully induced by the shape, size aspect ratio of the nanomaterial and formulation in addition to interfacial interactions between them. Besides, the reproducibility, reliability, toxicity, safeness of metal oxide nanomaterials and disposal mechanisms are decision criteria for its genuine prospection. More work needs to be done in order to explore this area of research as there are more of the application of metal oxide nanomaterials in biosensors than their electrochemical sensors counterpart. More work on the synergistic effect of bi/trimetallic metal oxide nanomaterial can be explored. To fully explore the synergistic effect of bi/trimetallic metal oxide nanomaterial, an advanced insight of the framework and mechanisms of their activities is imperative in more-in-depth study. It is very important to keep in mind accordingly that merely specialities cooperative task can result into groundbreaking research. This approach is highly needed in the scientific community across the globe; hence, the use of variety of metal oxide nanomaterials will uncover greater significant uses.

References

1. A.H. Kaksonen, N.J. Boxall, Y. Gumulya, H.N. Khaleque, C. Morris, T. Bohu, K.Y. Cheng, K.M. Usher, A.M. Lakaniemi, Recent progress in biohydrometallurgy and microbial characterisation. Hydrometallurgy **180**, 7–25 (2018). https://doi.org/10.1016/j.hydromet.2018.06.018
2. G. Aragay, A. Merkoçi, Nanomaterials application in electrochemical detection of heavy metals. Electrochim. Acta **84**, 49–61 (2012). https://doi.org/10.1016/j.electacta.2012.04.044

3. B. Rohrig, *Smartphones Could You Last a Day Without Your Cell Phone?*, (n.d.). https://www.acs.org/content/dam/acsorg/education/resources/highschool/chemmatters/archive/chemmatters-april2015-smartphones.pdf
4. B.D. Bolzan, Effect of heavy metals on living organisms. World Sci. News **5**, 26–34 (2014)
5. S.F. Zhou, X.J. Han, H.L. Fan, J. Huang, Y.Q. Liu, Enhanced electrochemical performance for sensing Pb(II) based on graphene oxide incorporated mesoporous $MnFe_2O_4$ nanocomposites. J. Alloys Compd. **747**, 447–454 (2018). https://doi.org/10.1016/j.jallcom.2018.03.037
6. B.K. Bansod, T. Kumar, R. Thakur, S. Rana, I. Singh, A review on various electrochemical techniques for heavy metal ions detection with different sensing platforms. Biosens. Bioelectron. **94**, 443–455 (2017). https://doi.org/10.1016/j.bios.2017.03.031
7. L. Cui, J. Wu, H. Ju, Electrochemical sensing of heavy metal ions with inorganic, organic and bio-materials. Biosens. Bioelectron. **63**, 276–286 (2015). https://doi.org/10.1016/j.bios.2014.07.052
8. P. Kumar, K.H. Kim, V. Bansal, T. Lazarides, N. Kumar, Progress in the sensing techniques for heavy metal ions using nanomaterials. J. Ind. Eng. Chem. **54**, 30–43 (2017). https://doi.org/10.1016/j.jiec.2017.06.010
9. J. Wang, *Analytical Electrochemistry*, 2nd edn. (Wiley, Hoboken, 2000)
10. A. Waheed, M. Mansha, N. Ullah, Nanomaterials-based electrochemical detection of heavy metals in water: current status, challenges and future direction. TrAC—Trends Anal. Chem. **105**, 37–51 (2018). https://doi.org/10.1016/j.trac.2018.04.012
11. S. Deshmukh, G. Kandasamy, R.K. Upadhyay, G. Bhattacharya, D. Banerjee, D. Maity, M.A. Deshusses, S.S. Roy, Terephthalic acid capped iron oxide nanoparticles for sensitive electrochemical detection of heavy metal ions in water. J. Electroanal. Chem. **788**, 91–98 (2017). https://doi.org/10.1016/j.jelechem.2017.01.064
12. S.H. Choi, J.P. Choi, Gold nanoparticle-based electrochemical sensor for the detection of toxic metal ions in water. J. Res. Environ. Earth Sci. **3**, 47–54 (2017)
13. I.A. Tayeb, K. Abdul Razak, Development of gold nanoparticles modified electrodes for the detection of heavy metal ions. J. Phys.: Conf. Ser. 1083 (2018). https://doi.org/10.1088/1742-6596/1083/1/012044
14. Y. Song, C. Bian, J. Tong, Y. Li, S. Xia, The graphene/L-cysteine/gold-modified electrode for the differential pulse stripping voltammetry detection of trace levels of cadmium. Micromachines **7**, 1–9 (2016). https://doi.org/10.3390/mi7060103
15. S.T. Palisoc, N.C.C. Valeza, M.T. Natividad, Fabrication of an effective gold nanoparticle/graphene/nafion® modified glassy carbon electrode for high sensitive detection of trace Cd^{2+}, Pb^{2+} and Cu^{2+} in tobacco and tobacco products. Int. J. Electrochem. Sci. **12**, 3859–3872 (2017). https://doi.org/10.20964/2017.05.14
16. X. Zhu, J. Tong, C. Bian, C. Gao, S. Xia, The polypyrrole/multiwalled carbon nanotube modified Au microelectrode for sensitive electrochemical detection of trace levels of Pb^{2+}. Micromachines 8 (2017). https://doi.org/10.3390/mi8030086
17. J. Gayathri, K.S. Selvan, S.S. Narayanan, Fabrication of carbon nanotube and synthesized Octadentate ligand modified electrode for determination of Hg(II) in Sea water and Lake water using square wave anodic stripping voltammetry. Sens. Bio-Sens. Res. **19**, 1–6 (2018). https://doi.org/10.1016/j.sbsr.2018.02.006
18. X. Xuan, J.Y. Park, A miniaturized and flexible cadmium and lead ion detection sensor based on micro-patterned reduced graphene oxide/carbon nanotube/bismuth composite electrodes. Sens. Actuators, B: Chem. **255**, 1220–1227 (2018). https://doi.org/10.1016/j.snb.2017.08.046
19. S. Lee, J. Oh, D. Kim, Y. Piao, A sensitive electrochemical sensor using an iron oxide/graphene composite for the simultaneous detection of heavy metal ions. Talanta **160**, 528–536 (2016). https://doi.org/10.1016/j.talanta.2016.07.034
20. Y. Ma, Y. Wang, D. Xie, Y. Gu, X. Zhu, H. Zhang, G. Wang, Y. Zhang, H. Zhao, Hierarchical MgFe-layered double hydroxide microsphere/graphene composite for simultaneous electrochemical determination of trace Pb(II) and Cd(II). Chem. Eng. J. **347**, 953–962 (2018). https://doi.org/10.1016/j.cej.2018.04.172

21. X.Y. Yu, Z.G. Liu, X.J. Huang, Nanostructured metal oxides/hydroxides-based electrochemical sensor for monitoring environmental micropollutants. Trends Environ. Anal. Chem. **3–4**, 28–35 (2014). https://doi.org/10.1016/j.teac.2014.07.001

22. S. Kempahanumakkagari, A. Deep, K.-H. Kim, S. Kumar Kailasa, H.-O. Yoon, Nanomaterial-based electrochemical sensors for arsenic—A review. Biosens. Bioelectr. **95**, 106–116 (2017). https://doi.org/10.1016/j.bios.2017.04.013

23. S.R.V.S. Prasanna, K. Balaji, S. Pandey, S. Rana, *Metal Oxide Based Nanomaterials and Their Polymer Nanocomposites* (Elsevier Inc., Amsterdam, 2019), pp. 123–144. https://doi.org/10.1016/b978-0-12-814615-6.00004-7

24. A. Mao, H. Li, Z. Cai, X. Hu, Determination of mercury using a glassy carbon electrode modified with nano TiO_2 and multi-walled carbon nanotubes composites dispersed in a novel cationic surfactant. J. Electroanal. Chem. **751**, 23–29 (2015). https://doi.org/10.1016/j.jelechem.2015.04.034

25. S. Ramezani, M. Ghobadi, B.N. Bideh, Voltammetric monitoring of Cd (II) by nano-TiO_2 modified carbon paste electrode sensitized using 1,2-bis-[o-aminophenyl thio] ethane as a new ion receptor. Sens. Actuators, B: Chem. **192**, 648–657 (2014). https://doi.org/10.1016/j.snb.2013.11.033

26. F.M. Liu, Y. Zhang, W. Yin, C.J. Hou, D.Q. Huo, B. He, L.L. Qian, H.B. Fa, A high–selectivity electrochemical sensor for ultra-trace lead (II) detection based on a nanocomposite consisting of nitrogen-doped graphene/gold nanoparticles functionalized with ETBD and Fe_3O_4–TiO_2 core—shell nanoparticles. Sens. Actuators, B: Chem. **242**, 889–896 (2017). https://doi.org/10.1016/j.snb.2016.09.167

27. Y. Wei, R. Yang, X.Y. Yu, L. Wang, J.H. Liu, X.J. Huang, Stripping voltammetry study of ultra-trace toxic metal ions on highly selectively adsorptive porous magnesium oxide nanoflowers. Analyst **137**, 2183–2191 (2012). https://doi.org/10.1039/c2an15939b

28. S.S. Li, W.Y. Zhou, M. Jiang, L.N. Li, Y.F. Sun, Z. Guo, J.H. Liu, X.J. Huang, Insights into diverse performance for the electroanalysis of Pb(II) on Fe_2O_3 nanorods and hollow nanocubes: toward analysis of adsorption sites. Electrochim. Acta **288**, 42–51 (2018). https://doi.org/10.1016/j.electacta.2018.08.069

29. S. Xiong, B. Yang, D. Cai, G. Qiu, Z. Wu, Individual and simultaneous stripping voltammetric and mutual interference analysis of Cd^{2+}, Pb^{2+} and Hg^{2+} with reduced graphene oxide-Fe_3O_4 nanocomposites. Electrochim. Acta **185**, 52–61 (2015). https://doi.org/10.1016/j.electacta.2015.10.114

30. Y. Sun, W. Zhang, H. Yu, C. Hou, D.S. Li, Y. Zhang, Y. Liu, Controlled synthesis various shapes Fe_3O_4 decorated reduced graphene oxide applied in the electrochemical detection. J. Alloy. Compd. **638**, 182–187 (2015). https://doi.org/10.1016/j.jallcom.2015.03.061

31. G. Bhanjana, N. Dilbaghi, R. Kumar, A. Umar, S. Kumar, SnO_2 quantum dots as novel platform for electrochemical sensing of cadmium. Electrochim. Acta **169**, 97–102 (2015). https://doi.org/10.1016/j.electacta.2015.04.045

32. M. Yang, T.J. Jiang, Z. Guo, J.H. Liu, Y.F. Sun, X. Chen, X.J. Huang, Sensitivity and selectivity sensing cadmium(II) using amination functionalized porous SnO_2 nanowire bundles-room temperature ionic liquid nanocomposite: combined efficient cation capture with control experimental conditions. Sens. Actuators B: Chem. **240**, 887–894 (2017). https://doi.org/10.1016/j.snb.2016.09.060

33. X. Cui, X. Fang, H. Zhao, Z. Li, H. Ren, Fabrication of thiazole derivatives functionalized graphene decorated with fluorine, chlorine and iodine@SnO_2 nanoparticles for highly sensitive detection of heavy metal ions. Colloids Surf. A **546**, 153–162 (2018). https://doi.org/10.1016/j.colsurfa.2018.03.004

34. L. Yuan-Yuan, C. Meng-Ni, G. Yi-Li, Y. Jian-Mao, M.A. Xiao-Yu, L. Jian-Yun, Preparation of zinc oxide-graphene composite modified electrodes for detection of trace Pb(II). Chinese J. Anal. Chem. **43**, 1395–1401 (2015). https://doi.org/10.1016/s1872-2040(15)60862-3

35. J. Yukird, P. Kongsittikul, J. Qin, O. Chailapakul, N. Rodthongkum, ZnO@graphene nanocomposite modified electrode for sensitive and simultaneous detection of Cd (II) and Pb (II). Synth. Met. **245**, 251–259 (2018). https://doi.org/10.1016/j.synthmet.2018.09.012

36. A. Salimi, B. Pourbahram, S. Mansouri-Majd, R. Hallaj, Manganese oxide nanoflakes/multi-walled carbon nanotubes/chitosan nanocomposite modified glassy carbon electrode as a novel electrochemical sensor for chromium (III) detection. Electrochim. Acta **156**, 207–215 (2015). https://doi.org/10.1016/j.electacta.2014.12.146
37. M. Fayazi, M.A. Taher, D. Afzali, A. Mostafavi, Fe_3O_4 and MnO_2 assembled on halloysite nanotubes: a highly efficient solid-phase extractant for electrochemical detection of mercury(II) ions. Sens. Actuators, B: Chem. **228**, 1–9 (2016). https://doi.org/10.1016/j.snb.2015.12.107
38. Y. Tang, S. Deng, L. Ye, C. Yang, Q. Yuan, J. Zhang, C. Zhao, Effects of unfolded and intercalated halloysites on mechanical properties of halloysite-epoxy nanocomposites. Composites: Part A **42**, 345–354 (2011). https://doi.org/10.1016/j.compositesa.2010.12.003
39. Y. Xie, D. Qian, D. Wu, X. Ma, Magnetic halloysite nanotubes/iron oxide composites for the adsorption of dyes. Chem. Eng. J. **168**, 959–963 (2011). https://doi.org/10.1016/j.cej.2011.02.031

Chapter 8
Application of Metal Oxides Electrodes

Chikaodili Chukwuneke, Joshua O. Madu, Feyisayo V. Adams,
and Oluwagbenga T. Johnson

Abstract The search for engineering materials that can withstand the high demands of the emerging technologies in the fields of bio-engineering, aerospace engineering, medicine, environmental protection, renewable energy and manufacturing industries continues to thrive and find relevance in the today's world. Metal oxides-based electrodes possess exceptional properties which qualify them as suitable engineering materials with wide range of applications such as sensors, semiconductors, energy storage, lithium-ion batteries and solar cells. This paper focuses on the use of various metal oxide-based electrodes (metal oxide, transition metal oxide, mixed metal oxide, transition, and hybrid systems) and how they have improved certain parameters of energy storage such as life cycle, capacitance, nominal voltage in above mentioned application prospects. This paper describes the novel concept of lithium metal oxide electrode materials which are of value to researchers in developing high-energy and enhanced-cyclability electrochemical capacitors comparable to Li-ion batteries. In order to fully achieve the potential of metal oxide electrodes in the future, significant efforts need to be directed to producing low cost and environment-friendly materials.

8.1 Introduction

Metal oxides exhibits several properties ranging from semiconductors and insulators, electrical properties from metals, and they find applications in different areas such superconductors, magnets, sensors, medical devices and lighting [1]. Metal oxides

C. Chukwuneke · J. O. Madu · F. V. Adams
Department of Petroleum Chemistry, American University of Nigeria, Yola, Nigeria

F. V. Adams (✉) · O. T. Johnson
Department of Metallurgy, School of Mining, Metallurgy and Chemical Engineering,
Faculty of Engineering and the Built Environment, University of Johannesburg, P.O. Box 17011,
Doornfontein 2028, South Africa
e-mail: feyikayo@gmail.com

O. T. Johnson
Department of Mining and Metallurgical Engineering, University of Namibia, Ongwediva
Engineering Campus, Windhoek, Namibia

© Springer Nature Switzerland AG 2020
O. M. Ama and S. S. Ray (eds.), *Nanostructured Metal-Oxide Electrode
Materials for Water Purification*, Engineering Materials,
https://doi.org/10.1007/978-3-030-43346-8_8

have the capacity to change their electrical conductivity relatively to the composition of the surrounding atmosphere [2].

Metal oxide nanomaterials (MON) were introduced with a view to improving the essential attributes of metal oxides for additional applications. These MON have become a growing asset in several industries because of their physical, electrical and chemical properties compared to similar materials. The MON find application in medicine, environmental remediation, renewable energy, personal care and water treatment [1, 3].

Metal oxide semiconductor comprises of three components namely an insulating film, a substrate and a metal electrode. The metal electrode (upper layer) serves as the conductor, the middle layer which is an insulator could be a glass or silicon dioxide, and the third layer, which is the substrate is also a conductive layer comprises of crystal silicon. This third layer can serve as a semiconductor whose conductivity can be varied thermally and doping.

Metal oxides are used as electrode materials for a electrocatalysis and electrosynthesis [4]. An oxide single crystal is an ideal oxide electrode for a major kinetic study; in this study, the electrode's surface that is well-defined chemically, geometrically and morphologically is exposed to the electrolyte [4, 5]. Metal oxide with better catalytic activity and considerable surface area are often used as electrodes. Most metal oxides can undergo proton-insertion reactions, these metal oxides find application in batteries and are currently being cited as potential electrochromic materials [6]. The oxygen electrode play significant role in conversion, storage and conservation of energy.

Electrodes are chemical materials that are of significant economic importance for the digital era in which we live. They can dictate how much energy is available for our gadgets, our cars, homes, offices, serve as sensors as well as aid the selective purification and synthesis of modern materials. Metal oxide electrodes do not fall short of this expertise instead; they have improved upon the repertoire when placed in tandem with prior materials used in their stead.

Metal oxide-based electrodes have left their footprint in the fabrication of energy storage devices (super-capacitors and batteries), photovoltaic solar cells (DSSC's), and renewable energy production plants. The focus of this review is to highlight the use of variant metal oxide-based electrodes (metal oxide, transition metal oxide, mixed transition metal oxide, and hybrid systems) and show how they have improved certain parameters/markers of better energy storage such as life cycle, capacitance, nominal voltage in aforesaid application prospects.

When considering selection of metal oxides for electrode in a particular application, the nanostructures of the metal oxides are of paramount importance. The significance of the metal oxide nanostructures as electrode in several application is as a result of their ability to provide large surface area to volume ratio. Therefore, is responsible for carrying high number of the membrane molecules and allow enormous surface area for the oxidation of the analyte molecules on the surface of the electrode [7]. Secondly, they provide biocompatibility on the electrode's surface and they enable easy coating of different membrane on the surface of the electrode. They

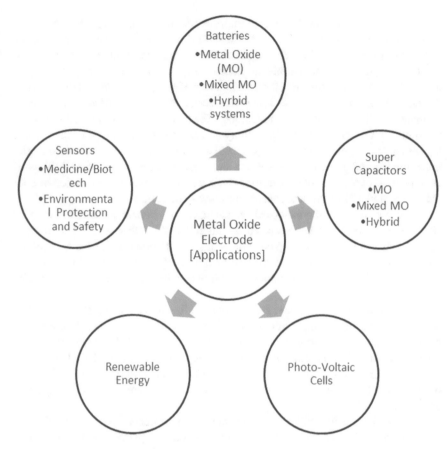

Fig. 8.1 Applications of metal oxides electrodes

also provide rapid electron transfer, which catalyzed slow electrode reactions (redox-active). Metal oxide electrodes are applied in several areas, and their applications are vast, but this work will be streamlined to the application of metal oxide electrodes in medicine, as super-capacitors, batteries, photo-voltaic cells and sensor (Fig. 8.1).

8.1.1 Metal Oxide Electrodes as Sensors in Medicine and Other Areas

Sensors are scientific response to a data driven world which demands that collation of raw fact for an innumerable parameter of interest is done with pitch perfect precision and accuracy. They are widely used in bio-engineering, environmental protection, aerospace engineering, manufacturing industries, and medicine [8–11].

The primary goal of medicine is to improve life expectancy through early detection, prevention and curing of diseases. Biosensors for blood glucose by metal oxide-based electrodes have enhanced early detection of diabetes mellitus which has had a deleterious effect on human health. There are quite a number of analytical techniques that have been utilized in the development of glucose sensors and they include fluorescent spectroscopy, conductometric, electrochemical, optical and colorimetric methods [12–16]. Among these techniques, the electrochemical method has taken the spotlight due to its superior selectivity and sensitivity. By reason of their working principles, the electrochemical [glucose] sensors are divided into amperometric, impedimetric and potentiometric sensors and metal-oxide electrodes find their applications in all of these. Metal oxide electrodes can be used in the fabrication of both enzymatic and non-enzymatic sensors for glucose detection.

The use of metal oxide nanoparticles electrodes whether modified physically or chemically in medicine has attracted the attention researchers [17]. This interest has been due to the outstanding catalytic and physical properties, ease fabrication, biocompatibility, controllable shape and size as well as strong adsorption ability, electron transfer kinetics and chemical stability of these metal oxides nanoparticles [17].

Metal oxide semiconductor sensors have been found to be the most cost effective and require low-power applications such as disposable medical. In 2010, Rahman et al. [18] reported several work on the nanostructured metal oxides for glucose sensors. Metal oxide nanomaterials such as copper oxide (CuO) [19–21], zirconia (ZrO_2) [22, 23], titanium dioxide (TiO_2) [24–26], cerium oxide (CeO_2) [27, 28], iron oxide (Fe_3O_4) [29], and zinc oxide (ZnO) [30, 31] are commonly used in glucose detection.

Among the metal oxides that appealed to researchers for the fabrication of electrodes is MnO_2 nanoparticles/chitosan-modified pencil graphite electrode (MnO_2NPs/CS/PGE). This electrode was used for selective and sensitive determination of furosemide (FUR). It was discovered that under the optimized experimental conditions, the modified electrode (g-MnO_2/CS/PGE) provided a linear response over the concentration range of 0.05–4.20 mmol/L FUR with a low limit of detection of 4.44 nmol/L for the 1st peak and 3.88 nmol/L for the 2nd peak [32]. The electrodes were tested for reproducibility and selectivity, and the results were good. The electrodes were also applied in spiked urine samples and they showed good precision and accuracy [32]. In another investigation, the application of metal oxide electrodes as sensor in medicine was carried out using cyclic voltammetry, EIS and square wave voltammetry techniques in 0.1 M phosphate buffer solution PBS at pH 7 [33]. Nanotubes doped with Nikel oxide, Zinc oxide and Iron oxide nanoparticles were used as electrodes. The reactivity of the metal oxide electrodes was greatly improved by modification with nanoparticles. A good linear property in the concentration range from 4×10^{-5} µM to 6.25 µM was obtained from the quantitative analysis of dopamine (DA) with a limit of detection of at least 3.742×10^{-7} M and a maximum of 1.386×10^{-6} [33]. The modified metal oxide electrodes showed great stability and long shelf life under ambient conditions.

A synopsis of various metal-oxide electrodes that are utilized in glucose detection in terms of their electrode matrix, detection technique, presence of enzymatic/non-enzymatic mode of operations, detection limit, response time/applied potential (V), and the referenced work have been reported previously [18].

Also biosensors based on metal oxide electrodes have been developed in order to study Parkinson's disease, a common neurodegenerative ailment. Parkinson's disease is birthed out of the death of dopamine bearing neurons in the substantia nigra; a part of the mid-brain [34]. The amino-acid tyrosine serves as an important starting material for the synthesis of levodopa which is also an essential precursor in the production of dopamine in the human body. Since dopamine lacks the capacity to invade the blood-brain barrier, but levodopa can, the former cannot be administered directly and hence levodopa is used in the treatment of Parkinson's disease [35]. In order to comprehend the treatment and elucidate the mechanism of Parkinson's disease, levodopa and tyrosine are the biomarkers of interest. Beitollahi and Fariba exploited this idea in developing a novel GO/ZnO nano rods nano composite screen printed electrode aimed at simultaneous detection of tyrosine and levodopa. The miniature measuring device was tested with human blood serum and urine samples and it exhibited a remarkable recovery of >98% and >97.5% for levodopa and tyrosine in blood serum, and also >97.6% and >98% for levodopa and tyrosine in urine respectively [36].

8.1.2 Environmental Protection and Safety

Safety is an important global consideration in the design of all sorts of systems, and sensors serve as checks to curb untoward events to resources and human health. Gases such as CO_x, NO_x, H_2, Phenols and Quinone's are of industrial and environmental significance; hence the need to keep track of their abundance. Hydrogen is a flammable, colorless and odorless gas with immense industrial and domestic applications such as in hardening of oils, gas cylinders in laboratories, and hydrogen-powered vehicles. The aforementioned properties of hydrogen demand the need to stalk its footprint so as to ensure safety where it is used. A plethora of Yttra Stabilized Zirconia (YSZ)-based and potentiometric sensors make use of metal oxide sensing electrode in the detection of hydrogen [37–40]. The metal oxide electrode (ITO) serves as the sensing electrode (SE). This sensor works by reason of the kinetic balance of an electrochemical redox reaction—using an air/hydrogen mixture, which aids the electrodes towards achieving steady state. Below are the underlying redox reactions governing the working principle of the sensors [37].

$$\text{Cathode: } 2O_2 + 2e^- \rightleftharpoons O^{2-}$$
$$\text{Anode: } H_2 + O_{2-} \rightleftharpoons H_2O + 2e^-$$

Table 8.1 A summary of metal-oxide based sensors with a bias for hydrogen detection [46]

Sensors	Oxide type
CuO–ZnO–Al$_2$O$_3$/Pt| YSZ|Pt	Binary
ZnO|YSZ|Pt	Binary
ITO|YSZ|Ag	Binary
ITO|YSZ|Pt	Binary
ITO|YSZ|Pt	Binary
SnO$_2$(+YSZ)|YSZ|NiO–TiO$_2$	Binary
Cr$_2$O$_3$/Al$_2$O$_3$/SnO$_2$(+YSZ)|YSZ|Pt	Binary
ZnO–Zn$_3$Ta$_2$O$_8$|YSZ|Pt	Binary/Ternary
ZnO+(+Ta$_2$O$_5$)|YSZ|Pt	Binary

Lu and co-authors [38] carried out a comparative study on the synthesis of high temperature hydrogen sensors based on the structure MO$_x$|YSZ|Pt, where MO$_x$ represents metal oxide and every other thing retains their usual meaning. The study showed that on testing various metal-oxides such as ZnO, SnO$_2$, In$_2$O$_3$, WO$_3$, TiO$_2$, CuO, Fe$_2$O$_3$, Mn$_3$O$_4$, Co$_3$O$_4$, Cr$_2$O$_3$, NiO for sensing electrode prospect, ZnO at elevated temperatures of 450–600 °C proved to have the highest sensitivity of 50–500 ppm of H$_2$. The output of that study stresses clearly, the significance of metal-oxide electrodes in hydrogen sensor design [38]. Table 8.1 shows other YSZ sensors that harness the conductivity of metal oxides in making sensors. Other metal oxides such as CdWO$_4$, ZnWO$_4$, MnWO$_4$, LaMnO$_3$, ZnTa$_2$O$_6$, ZnO–Zn$_3$Ta$_2$O$_8$, ZnO+(+Ta$_2$O$_5$), Fe doped (La, Sr)CrO$_{3-\delta}$ perovskite oxide, have been developed as sensing electrodes in the design of hydrogen sensors [39, 41–45].

CO$_x$ and NO$_x$ gases are products of fossil fuel combustion and are well known greenhouse gases. They not only deplete the ozone layer, serve as precursors for acid rain, but are air pollutants that can combine with the haemoglobin in the blood and form complexes that are responsible for ailments ranging from simple headaches to respiratory diseases and to acute consequences such as death. The said effects require that their concentration in the environment need be monitored so as to mitigate their realization. Without a doubt, metal-oxide electrodes have been influential in the production of CO$_x$ and NO$_x$ sensors as it has been reported by many research groups. In tackling the problem of thermal and chemical instability exhibited by oxy-acids used in tandem with solid electrolyte, Jong Won Yoon and his team [47] made use of LaFeO$_3$ a semiconductor perovskite type oxide as replacement materials for the sensing electrodes. This was to enable sensor applications at high temperatures (400 and 450 °C) for NO$_x$ monitoring in exhausts fumes. The result of the study showed that aside the improvement in chemical and thermal stability, the sensor with the structure LaFeO$_3$/YSZ/Pt had a stable and reproducible EMF and a rapid response time of 4 min at 400 °C and 3.3 min at 450 °C. The electrochemical reactions that were assumed to be occurring at the electrodes are given below.

$$\text{Cathode: } NO_2 + 2e^- \rightarrow NO + O^{2-}$$

$$\text{Anode: } 2O^{2-} \rightarrow O_2 + 4e^-$$

However, some automobile engines such as spark ignition engines have their exhaust gas eluting at temperatures higher than 450 °C. To resolve issues emanating from higher operating temperatures, Xiong and his fellow researchers [48] upped the ante by pushing the boundaries of detection and sensitivity of NO_2 to temperatures >700 °C for automobile applications. The team explored different spinel oxides: $NiFe_{0.5}Cr_{1.5}O_4$, $NiFe_{0.75}Cr_{1.25}O_4$, $NiFe_{1.25}Cr_{0.75}O_4$, $NiFe_{1.75}Cr_{0.25}O_4$, $NiFe_{1.9}Al_{0.1}O_4$, $NiO + NiCr_2O_4$, and $CuO + NiCr_2O_4$ as sensing electrodes of which $NiFe_{1.9}Al_{0.1}O_4$ electrode showed a high selectivity to NO_2 with a response time of 8 s at 703 and 740 °C. Other metal oxides such as Cr_2O_3–WO_3, $MnCr_2O_4$, $SmFeO_3$, CuO, La_2CuO_4 and even spinel type metal oxides such as $La_{0.8}Sr_{0.2}CrO_3$, CuO–$CuCr_2O_4$, $La_{0.6}Ca_{0.4}Mn_{1-x}Me_xO_3$ have been utilized in making sensing electrodes for sensor applications in NOx detection [46, 49] (Table 8.2).

In the development of CO_x mixed potential gas sensors, MO electrodes have been utilized in making both sensing electrodes (SE) and the reference electrode (RE). It has been reported that strontium-doped lanthanum manganite (LSM) has been used as a counter electrode in a screen printed mixed potential sensor [50]. The sensor was of the structure LSM/YSZ/Au and it showed sensitivity towards both CO and NO_2 though with a greater sensitivity (and hence selectivity) to CO. The redox reactions occurring at the electrodes are explained below:

$$O_2 + 4e^- \rightarrow 2O^{2-}$$
$$CO + O^{2-} \rightarrow CO_2 + 2e^-$$
$$NO_2 + 2e^- \rightarrow NO + O^{2-}$$

$ZnCr_2O_4$, a metal oxide electrode with a high sensitivity towards CO_x has been proposed by Fujio et al. [51]. The sensor is of the structure $ZnCr_2O_4(+M)|YSZ|ZnCr_2O_4$ where M = Precious metal. Various precious metals such as Au, Ag, Pd, Ir, Pt, Rh, Ru where used in composite with the metal oxide of which the $ZnCr_2O_4$ (+Au) turned out to produce the highest sensitivity. When exposed to 100 ppm of a representative sample of exhaust gases under humid conditions at 550 °C, the sensor—$ZnCr_2O_4(+Au)|YSZ|ZnCr_2O_4$—showed the highest sensitivity towards CO and a similar response was observed when subjected to 800 ppm of CO.

Other metal oxides such as Nb_2O_5, $Zn_2SnO_4 + SnO_2$, TiO_2, have been utilized in the manufacturing of CO_x sensitive sensors [46]. Table 8.3 shows a summary of CO sensors using metal oxide electrodes either as a SE or RE or both.

Industrialization has necessitated the application of phenolic compounds in the production of banal commodities such as papers, plastics, pesticides, paints, and drugs. However, their toxic nature to human health and the ecosystem places necessity on the need for their close inspection. Sensors harnessing the chemical sensing capacity of metal-oxides have been used in making electrodes to serve the foresaid purpose. A carbon powder modified with PbO_2 composite electrode was designed by

Table 8.2 A summary of metal-oxide based sensors with a bias for NO_x detection [46]

Sensors	Oxide type
$CdMn_2O_4$|YSZ|Pt	Ternary
$CdCr_2O_4$|YSZ|Pt	Ternary
WO_3|YSZ|Pt	Binary
$NiCr_2O_4$|YSZ|Pt	Ternary
$ZnCr_2O_4$|YSZ|Pt	Ternary
$ZnFe_2O_4$|YSZ|Pt	Ternary
WO_3/Pt (or Au)|YSZ|Pt (or Au)	Binary
$V_2O_5(+Al_2O_3)$|YSZ|Pt	Binary
Cr_2O_3|YSZ|Pt	Binary
Cr_2O_3|YSZ|Pt	Binary
Cr_2O_3|YSZ|Pt	Binary
$La_{0.8}Sr_{0.2}FeO_3$|YSZ|Pt	Quaternary
ZnO|YSZ|Pt	Binary
$LaFeO_3$|YSZ|Pt	Ternary
NiO|YSZ|Pt	Binary
$La_{0.8}Sr_{0.2}CrO_3$|YSZ|Pt	Quatenary
$CuO–CuCr_2O_4$|ScSZ|Pt	Binary/Ternary
NiO(+Au)|YSZ|Pt	Binary
$La_{0.6}Ca_{0.4}Mn_{1-x}Me_xO_3$|YSZ|Pt	
NiO(+YSZ)|YSZ|Pt	Binary
NiO(+Cr)|YSZ|Pt	Binary
$Ni_{1-x}Co_xO$|YSZ|Pt	Ternary
La_2CuO_4|YSZ|Pt	Ternary
CuO|YSZ|Pt	Binary
$SmFeO_3$|YSZ|Pt	Ternary
$MnCr_2O_4$|YSZ|Pt	Ternary
$WO_3–Cr_2O_3$|YSZ|Pt	Binary

Sljuki et al. [52] in detecting the presence of phenolic compounds using phenol, p-nitrophenol, and 2-chlorophenol as test materials. The outcome of the study showed a detection limit of 0.19 μM promising results for the detection of various phenols in high acidity and high saline waste water effluents. Owing to the reaction-promoting nature of CeO_2 based composite, and also the stability, cost-effective and electron mediating capacity of ZnO, Singh et al. [53], synthesized a crystalline CeO_2–ZnO composite Nano ellipsoids for the fabrication of 4-nitrophenol sensors. The fabricated sensor showed a reasonable detection limit of 1.163 μM and a sensitivity of ~0.120 μA/nM cm^2. The suggested electrochemical reaction, which governs the reaction occurring at the electrodes are given below [53, 54] (Fig. 8.2).

Table 8.3 A summary of metal-oxide based sensors with a bias for CO detection [46]

Sensors	Oxide type
Al_2O_3(+Pt)/Pt|YSZ|Pt	Binary
Au/Ba$_{0.9}$Gd$_{0.1}$SnO$_3$|YSZ|Pt	Quaternary
CuO–ZnO/Pt|YSZ|Pt	Binary
RuO_2|YSZ|Pt	Binary
SnO_2|YSZ|CdO	Ternary
Au/LaCoO$_3$|YSZ|Au	Ternary
LaMnO$_3$|YSZ|Y$_{0.17}$Tb$_{0.17}$Zr$_{0.66}$O$_{2-x}$	Ternary/Quaternary
LaMnO$_3$|YSZ|Y$_{0.17}$Tb$_{0.17}$Zr$_{0.66}$O$_{2-x}$	Ternary/Quaternary
Au–Co$_3$O$_4$|YSZ|Pt	Ternary
Nb$_2$O$_5$/Au|YSZ|Pt	Binary
LSM|YSZ|Au	Quaternary
TiO$_2$–Y$_2$O$_3$–Pd|YSZ|TiO$_2$	Binary
ZnCr$_2$O$_4$(+Au)|YSZ|ZnCr$_2$O$_4$	Ternary
Nb$_2$O$_5$(+Au)|YSZ|NiO(+Au)	Binary
Zn$_2$SnO$_4$(+SnO$_2$)|YSZ|Pt	Binary/Ternary

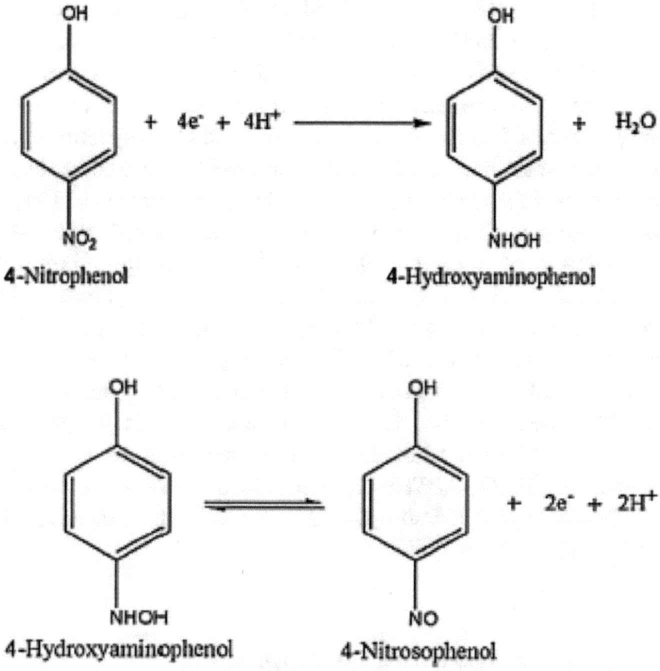

Fig. 8.2 Redox reaction occurring at the Electrodes in the detection of nitro-phenols

With the growing demand to make sensors and electrochemical analytical system less cumbersome Bi_2O_3 [55] has been used in designing electrodes for use in screen printed disposable sensors. The sensor was applied in a cyclic voltammogram detection system which was used to identify 2-nitrophenol, 4-nitrophenol and 2, 4-dinitrophenol in water samples. Other metal oxides such as Ag_2O, α-MnO_2, CuO nanotubes and Mn_2O_3–ZnO, nanoparticles have also been used as electrode materials in the fabrication of phenolic sensors [56–60].

Hydroquinone (HQ) and Catechol (CC) are another class of phenolic contaminants which have an enduring footprint on the environment (soil and water) and have been reported to cause respiratory problems, tinnitus, skin irritation and edema of internal organs [61]. MnO_2 has been explored in fabricating electrodes for HQ and CC detection. Prathap et al. [62] synthesized a PANI/MnO_2 nano composite electrode that was used in a differential pulse voltameter for the concurrent detection of HQ and CC. The electrodes exhibited a good response to HQ and CC with a detection limit of 0.13 and 0.16 μM and sensitivity of 0.8 and 0.5 $\mu A/\mu M$ respectively. Also, MnO_2 has been combined with a mesoporous graphene oxide as a nano composite electrode for a voltametric sensor aimed at HQ and CC [63]. The modified electrode proved to have a good selectivity and sensitivity to the analyte of interest in that, when it was subjected to a mixture of a 100-fold of SO_3^{2-}, Ni^{2+}, Ca^{2+}, S^{2-}, Fe^{3+}, Br^-, glucose, ascorbic acid, uric acid, 1000-fold, Cu^{2+}, NH_4^+, K^+, Mg^{2+}, Zn^{2+}, Na^+, NO_3^-, Cl^-, SO_4^{2-}, 5-fold resorcinol, and 0.1 μM CC and HQ, no interference was observed. But when it was tested with 3-fold phenols and nitro phenols, there was interference in the detection of HQ and CC. The sensor with the structure GO-mesoporous MnO_2/GCE has a detection limit of 7 nM and 10 nM for HQ and CC respectively. ZnO based nano materials have also been explored by a number of researchers for developing HQ sensors. Fe doped ZnO nano rods were fabricated by Umar et al. [64] to serve as electrodes in screen printed HQ sensors. The assembled Fe-ZnO nano rod electrode displayed a sensitivity of 18.60 μA mM^{-1} cm^{-2} and a detection limit of 0.51 μM. While Ahmad Umar et al. utilized ZnO for screen printed electrodes, Ameen and his team [65] employed same in synthesizing a ZnO/GCE composite electrode for p-HQ detection. The modified ZnO/GCE didn't reproduce the results of Fe–ZnO electrodes as it exhibited a sensitivity of 99.2 μA μM^{-1} cm^{-2} and a detection limit of 4.5 μM. There are also other studies, in which electrode designs for HQ and CC sensors using modified metal oxide electrodes such as Pt–MnO_2, TiO_2–SiC(HQ and BPA detection), Fe_3O_4-APTES-GO with electrode structure GCE/Pt–MnO_2, Pd@TiO_2–SiC/GCE, and AuNPs/Fe_3O_4-APTES-GO/GCE, respectively have been implemented [66–68] (Fig. 8.3).

8.1.3 Metal Oxide Electrodes as Semiconductors

Metal oxide electrodes have shown satisfactory capacity as semiconductors in various applications. They are used as photo anode materials in photo electrochemical reactions. Tungsten trioxide is a metal oxide electrode (n-type semiconductor)

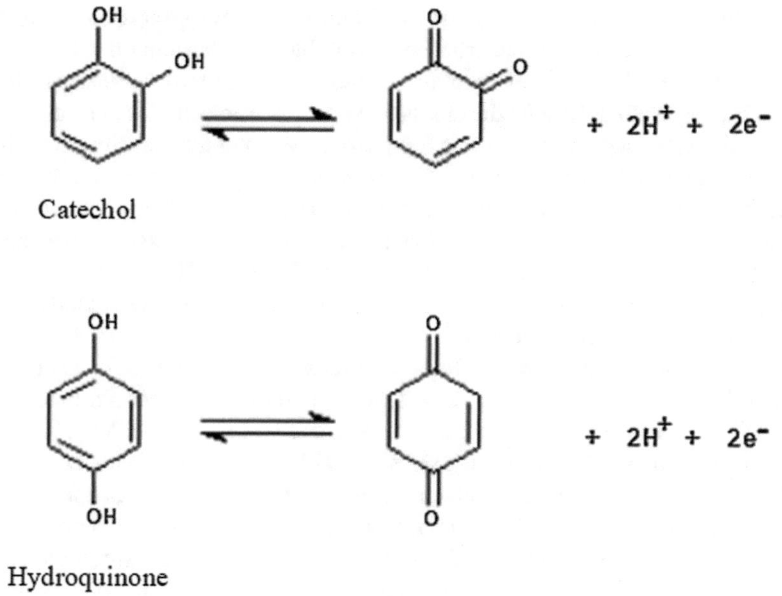

Catechol

Hydroquinone

Fig. 8.3 Electrochemical reaction governing the activities at the electrode

used as a photo anode; it has a high tendency to absorb visible spectrum [69]. Titanium dioxide (TiO_2), which is a great photocatalyst in several processes [70], it is a good material for solar hydrogen generation (H_2) and also used for solar water splitting [71]. A review was particularly carried out on the application of TiO_2 as semiconductor [72]. TiO_2 application in agriculture includes plant germination and growth, water purification, crop disease control, pesticides' degradation and pesticides residue detection [73]. Another metal oxide electrode used in the solar splitting of water is $-\alpha\text{-}Fe_2O_3$. However, some limitations were observed in its application; one of which is inadequate conduction band to reach appropriate efficiency [73].

Metal oxides have been used to modify organic semiconductors in order to improve their conductivities, band gap as well as their photo generated charge carriers. This application was as a result of metal oxides' exceptional optical and electrical properties.

Molybdenum trioxide (MoO_3) coupled to silver (Ag) has been used to upgrade the power conversion of semi-transparent plastic solar cells giving rise to enhanced performance [74]. Various power conversion efficiency (4.5% with a maximum visible region transparency of approximately 50–9.1% and a maximum visible region transparency of 5%) was achieved with the modified electrode by changing the thickness of the Ag layer in the metal oxide [74].

Super-capacitors are considered the electrochemical energy storage cells of the future. Their longevity, high power density, and high charge/discharge rates has placed them on the front burner as possible replacement materials for energy storage devices in handy digital assistants and mobile electronics, hybrid and electric automobiles. Super-capacitors is classified into three main groups based on their energy storage mechanisms; electrochemical double layer capacitor (EDLC), (a carbon based electrode), Pseudo-capacitor, (a redox metal or redox polymer electrode) and hybrid capacitors. Despite the merits of carbon based electrodes of electrochemical double layer capacitor such as a long service life and good cycling stability, it suffers a flaw of low energy density and low specific capacitance. Metal oxides are therefore used as vital tools in augmenting the specific capacitance and the energy density of super-capacitors [75]. Various oxides of transition metals such as Ruthenium (IV) oxide, Tungsten (VI) oxide, Copper (II) oxide, Zinc oxide, Nickel (II) oxide have been studied for a variety of applications to improve the capacitance and energy density of super-capacitors

A supercapacitor that consist of RuO_2 electrode was been reported by Patake et al. [76] to show a maximum specific capacitance of 650 Fg^{-1}, this was attributed to the surface treatments that impacted positively on the surface morphology. Table 8.4 showing a summary of some metal-oxides and their applications to super capacitors.

In addition to the pseudo capacitors, synergistic combinations of metal-oxides and carbon materials have been exploited for applications in hybrid super capacitors. Such hybrid systems permit the maximization of the high energy density, high specific capacitance of metal oxides in combination with high power density, rate capability, and stability of carbon nanomaterial. Asymmetric hybrid supercapacitors comprising of a metal oxide nanowire/single-walled carbon nanotube was synthesized by Chen et al. [86]. The outcome of their study showed that the hybrid system had a capacitance of 184 (Fg^{-1}), and energy density of 25.5 Wh/Kg.

Table 8.4 Examples of metal-oxide specific capacitor, specific capacitance and oxide type

Metal-oxide	Specific capacitance (Fg^{-1})	Oxide type	References
RuO_2	650	Binary	[76]
MnO_2	241	Binary	[77]
NiO	167	Binary	[78]
Co_3O_4	~1100	Binary	[79]
SnO_2	187	Binary	[80]
CuO	>62	Binary	[81]
TiO_2	62.8, 225	Binary	[82]
ZnO_2	~400	Binary	[83]
WO_3	588	Binary	[84]
$Na_4Mn_9O_{18}$	200	Ternary	[85]

8.1.4 Application of Metal Oxide-Based Electrode in Energy Storage

Today's high reliance on electrical energy powered device have placed pressure on large-scale energy storage and fast power energy delivery systems. These high demands are seen in different devices used for comfort and convenience in our daily life; from microchips to cars, cameras, implantable medical devices, computers and other communication devices. These pressures have encouraged research investigations that focus on increasing energy storing capacity and improved cycle life for energy storage devices [87–89].

There are many energy storage technologies in use however, electrochemical energy storage has been the most promising means of accumulating electricity in large-scale owing to its flexibility, energy conversion efficiency and easy maintenance [12]. Major electrochemical devices for electrical energy storage today are batteries and supercapacitors. These devices have attracted great interest and are essential as the leading power sources for portable electronics. They have the potentials to provide electricity for high energy demanding devices like electric vehicles and useful in other large-scale electrical devices [90].

The major challenge in advancing the use of these energy storage devices is in finding suitable electrode materials. The choice of electrode materials determines the electrode application, the suitability of several metal oxides as suitable electrode materials for different electronic devices and large energy storage devices have been investigated [91]. Each type of electrode material has its own merits and limitations. This have stimulated attempts to develop novel composite electrode materials via the coupling of materials that overcome the limitations of each other [92]. Getting suitable electrode materials that are durable with improved cycle life, cheap, readily available, capable of achieving high performance and meeting the needs for large storage devices have been the focus in electrochemical energy conversion and storage devices [92–95].

Studies on designs as well as the nature of material used in electrochemical energy systems have provided effective way to generating high-performance electrode and enhance the efficiency of electrochemical energy-storage devices in different application [96]. These efforts have culminated into the development of novel electrode materials together with advanced electrode architecture. Materials such as metal oxide in combination with graphene and nanomaterials are currently being investigated and utilized to achieve large scale energy conversion [97–99]. This have led to considerable progress in the preparation of commercial composites that are widely used in energy conversion and storage [98, 100].

Electrochemical energy storage systems consist of three essential parts, the negative electrode (anode), positive electrode (cathode) and electrolyte. These key components determine the performance and effectiveness of energy storage devices. Transition-metal-oxide including copper oxide, cobalt oxide, manganese oxides, iron oxides, tin oxides, vanadium oxides, molybdenum oxides, and iron oxides have been utilized as electrode materials for electrochemical energy storage devices [101–103].

The electrochemical capabilities, abundance in nature, low-cost and environmental friendliness underscores the research attention on metal oxides electrodes have drawn, occasioning the utilization of specific properties of these metal oxides in various applications. Manganese dioxide is projected among the most attractive metal oxide electrode materials for electrochemical energy storage applications. This is because of its high theoretical capacitance for storing electrical charge, being inexpensive, low toxicity, abundant natural reserves and eco-friendliness. Several forms of the oxides have been used as electrode materials in rechargeable lithium batteries, primary zinc-alkaline cells, primary lithium cells in aqueous electrolytes and electrodes of supercapacitors [104].

8.2   Application of Metal Oxide-Based Electrode in Lithium Ion Batteries

The nature of electrode materials have significant role in batteries performance and the need for improved batteries performance has never been more evident as batteries are essential to the power supply of many portable electronic devices used in today's society.

Lithium ion batteries (LIB) are among the most widely investigated rechargeable batteries. Its discovery is regarded as one of the most significant developments in energy storage for portable electronic devices. In the forms they exist today, lithium ion battery consist of a graphite anode (layered structure), a cathode made of lithium metal oxide ($LiMO_2$, e.g. $LiCoO_2$, $LiMO_2$, $LiNiO_2$ etc.) and lithium salt electrolyte (e.g. $LiPF_6$) in organic solvent (usually ethylene carbonate–dimethyl carbonate) imbedded in a separator [105]. LIB exhibits excellent performances in relation to energy density and are used in numerous electronics. However, the main challenge in the design of LIB is to ensure that the electrodes maintain their integrity over many discharge–recharge cycles [106]. More so, the commonly used graphite anode in Li-ion rechargeable batteries limits their capacity to store charge per unit weight. To overcome these limitations, scientist has vigorously sought for other anode materials such as metal oxides as substitute for graphite anodes. Metal oxides coupled with carbon materials (porous carbon, carbon nanotubes (CNTs) and graphene) are the favorite candidates. This combination has led to remarkably enhanced electrical conductivity, excellent mechanical and electrochemical stability as well as improved performance in terms of high reversible capacity. As electrode materials for LIBs hollow structure; SnO_2, Fe_2O_3, metal oxides displayed outstanding electrochemical efficiency due to the greatly enhanced diffusion kinetics and structural stability for lithium storage. These oxides are promising electrodes for lithium-ion batteries and operate through conversion reactions associated with much higher energy densities compared to intercalation reaction.

The use of conversion reaction metal oxides holds the promise of advanced Li-ion batteries with large capacity. Unfortunately, it suffers from several obstacles such as low conductivity, large volume changes and a large lithiation/de-lithiation voltage lag that give rise to low charge/discharge energy efficiency engendering severe battery heating during operation. These limitations have to be overcome before these promising electrodes can become a reality. A study by Polzot et al. [106], have shown that electrodes made of nanoparticles of transition-metal oxides [TMO], (where M is Fe, Co, Cu or Ni,) overcomes some of the above limitations and exhibit electrochemical capacities of 700 mA h g^{-1}, with hundred percent retention for up to 100 cycles. This nanoparticle TMO enabled high recharging rates and enhanced surface electrochemical reactivity. Likewise, titanium oxide, TO, with brookite structure have been investigated as anode material in LIB. This material offers advantages in cost, safety and environmental friendliness. Its maximum theoretical capacity was 335 mA h g^{-1}, equivalent to the insertion of one Li per TiO_2, (a complete reduction of $Ti^{4+} \rightarrow Ti^{3+}$). The morphology of titanium oxide greatly affects its electrochemical performance [105].

Chen and co-worker examined two nanomaterials systems that showed promising potentials as for use in LIB because their structure and design [107]. These materials are SnO_2 hollow spheres and anatase TiO_2 nanosheets with exposed high-energy facets. The authors revealed that the high lithium storage capacity, it availability in natural form (cassiterite), and modest intercalation potential of Li^+ in Sn makes SnO_2 hollow spheres an attractive anode material.

However, it suffers severe capacity fading due to enormous volume changes during alloying/dealloying; a challenge effectively mitigated by the use of hollow/mesoporous SnO_2 structures as the active anode material. This materials provides sufficient space to cushion the large internal stresses produced by huge volume changes. On the other hand, anatase TiO_2 has lower lithium storage capacity compared to the currently used graphite, but the use of nanostructures of both SnO_2 and TiO_2 created an enhanced storage for lithium [107].

Regarding cathodes materials, manganese-based compounds and olivine lithium metal phosphates are the main interest of many battery producers. These materials are readily accessible and environmentally friendly, lithium manganese spinel has been the most promising candidate; the manganese-based compounds seem almost perfect substitute for the high cost and somewhat noxious lithium cobalt oxide. Unfortunately, the use of lithium manganese spinel is restricted by availability of appropriate electrolyte as a result of the disintegration of manganese into the electrolyte upon cycling in lithium cells, especially above ambient temperatures [107]. Nickel cobalt manganese oxide, ($LiNi_{1/3}Co_{1/3}Mn_{1/3}O_2$) is another compound in the manganese family that has attracted the attention of researchers and manufacturers. This compound has extensively studied for usage in high energy and high power batteries, its advantages as cathode material include low cost, abundance and high performance [107].

The use of metal oxides in LIBs continues to be impeded long-term poor cycling stability and intrinsic low charge/ionic conductivity [12]. Researchers have tried to mitigate these problems through the manufacture of hollow structured mesoporous active materials of the metal oxides. The design of this materials generates short

diffusion path, high surface area and significantly diminished electrode pulveriza-
tion and polarization. This has given rise to exceptionally improved electrochemical
performance of LIB [108].

Wang and colleagues [12] investigated the effects of ordered mesoporous SnO_2
and found that mesoporous SnO_2 displays high cycling stability and high specific
capacity of up to 557 mA h g^{-1}, as well as high efficiency up 98.5%, even after 40
cycles at a high current density of 100 mA g^{-1}. This improvement in the electrochemical performance was ascribed to the ordered mesoporous structure of SnO_2
and large surface area that produced additional lithium storage sites and a large
electrode–electrolyte contact area for high Li^+ ions flux across the interface.

8.2.1 Metal Oxide Electrodes Applications in Solar Cells

Harvesting solar energy for the utilization in different electrochemical powered
device is a very attractive, yet a challenging task. Dye-sensitized solar cell (DSSC),
achieves efficient solar-to-electric power conversion through nanostructures. It is considered a potentially low cost alternative to traditional silicon-based photovoltaics
[110] and proper blending of TiO_2, SnO_2 or ZnO nanoparticles with ZnO nanotetrapods in DSSC photoanodes leads to greatly improved performance features of
DSSCs [113]. By exploiting material advantages of both ZnO nanotetrapods and
SnO_2 nanoparticles, Chen et al. [111] improved the energy conversion efficiencies
of flexible DSSC (>6%) using composite photoanodes incorporating ZnO and and
SnO_2 nanomaterials.

Different types of titanium dioxide (TiO_2) electrodes were used to study dye-
sensitized solar cells (DSSC). Chiba et al. [112] discovered that the ratio of incident
photon to current efficiency of DSSCs increased with increase in the haze of the
TiO_2 electrodes. This was intense at the region around the infrared wavelength. A
conversion efficiency of 11.1%, which indicated an improved conversion efficiency
was achieved in the course of the experiment.

The contact electrodes of photovoltaic cells considerably affect the light transmission and photovoltaic performance of the device. Hu and colleagues studied tin-oxide
(SnOx) on different designs of indium-tin-oxide (ITO)-free semitransparent bottom
electrode (SnOx/Ag or Cu/SnOx), and discovered that the metal efficiently pro-
tected against corrosion and simultaneously (SnOx) s as an electron extraction layer
which resulted in perovskite solar cells that are hysteresis-free with a stable power
conversion efficiency (PCE) of 15.3% and a remarkably high open circuit voltage
[113].

A semitransparent nanolayered metal/metal oxide electrode was used to enhance
the light transmission and power conversion efficiency for semitransparent PbS colloidal quantum dot solar cell. The nanolayered electrode increased the effectiveness
of light transmission in the visible region thereby enhancing the performance of the
photovoltaic cell by 28.6% and average visible transmittance by 59.6%, compared to

the standard Au film as the electrode. This demonstrates that nanolayered materials may provide an avenue for enhancing the device transparency and efficiency [114].

A sol–gel SnOx metal-oxide processed at room-temperature and Al:ZnO prepared at 100 °C were used to fuse wires together and also to "glue" them to substrates. This distinctive act resulted in a low sheet resistance (5.2 Ω sq^{-1}) and improved mean transmission of 87%. This idea permits transparent coatings even on temperature sensitive objects as it's highly robust and highly conductive. This concept has been used in transparent top-electrodes in efficient semitransparent organic solar cells [115].

8.3 Conclusion

A study on the application of metal oxides electrodes as sensors, semiconductors, energy storage, lithium-ion batteries and solar cells was carried out. Biosensors based on metal oxide electrodes used in medicine to detect diabetes mellitus, Parkinson's disease and other diseases that have deleterious effect on human health were discussed in detail. Moreover, a comprehensive study has also been done on factors that are affecting the sensitivity, selectivity and stability of the semiconductor metal oxide electrodes. The study establishes the novel concept of lithium metal oxide electrode materials which are of value to researchers in developing high-energy and enhanced-cyclability electrochemical capacitors comparable to Li-ion batteries. Further studies can be carried out to produce low cost, environment-friendly metal oxide electrodes for wider applications in the fields of bio-engineering, aerospace engineering, medicine, environmental protection, renewable energy and manufacturing.

References

1. A.K. Arora, V.S. Jaswal, K. Singh, R. Singh, Applications of metal/mixed metal oxides as photocatalyst: a review. Orient. J. Chem. **32**(4), 2035–2042 (2016)
2. A.I. Ayesh. Metal/metal-oxide nanoclusters for gas sensor applications. J. Nanomater. (2016). https://doi.org/10.1155/2016/2359019
3. J. Singh, T. Dutta, K.-H. Kim, M. Rawat, P. Samddar, P. Kumar, Green' synthesis of metals and their oxide nanoparticles: applications for environmental remediation. J. Nanobiotechnol. **16**(1), 84 (2018)
4. E. O'Sullivan, E.J. Calvo, Reactions at metal oxide electrodes, in *Comprehensive Chemical Kinetics, and Undefined 1988* (Elsevier, Amsterdam, 1988)
5. R. White, J. Bockris, B. Conway, E. Yeager, *Comprehensive Treatise of Electrochemistry. Vol. 8: Experimental Methods in Electrochemistry* (1984)
6. P. Sun, Z. Deng, P. Yang, X. Yu, Y. Chen, Z. Liang, H. Meng, W. Xie, S. Tan, W. Mai,."Freestanding CNT–WO3 hybrid electrodes for flexible asymmetric supercapacitors. J. Mater. Chem. A, **3**, 12076 (2015). pubs.rsc.org
7. S. Elhag, *Chemically Modified Metal Oxide Nanostructures Electrodes for Sensing and Energy Conversion* (2017)

8. F. Blais, Review of 20 years of range sensor development. J. Electron. Imaging (2004), spiedigitallibrary.org

9. P.S. Waggoner, H.G. Craighead, Micro- and nanomechanical sensors for environmental, chemical, and biological detection. Lab. Chip. **7**(10), 1238–1255 (2007)

10. D. Grieshaber, R. Mackenzie, J. Voros, E. Reimhult, Electrochemical biosensors—sensor principles and architectures. Kunststoffe Int. **8**(3), 1400–1458 (2008)

11. J. Homola, Surface plasmon resonance sensors for detection of chemical and biological species. Chem. Rev. **108**(2), 462–493 (2008)

12. Y. Wang, H. Xu, J. Zhang, G. Li, Electrochemical sensors for clinic analysis. Sensors **8**(4), 2043–2081 (2008)

13. M.A. Morikawa, N. Kimizuka, M. Yoshihara, T. Endo, New colorimetric detection of glucose by means of electron-accepting indicators: ligand substitution of [Fe(acac)3-n(phen)n]n + complexes triggered by electron transfer from glucose oxidase. Chem.—A Eur. J. **8**(24), 5580–5584 (2002)

14. Y. Miwa, M. Nishizawa, T. Matsue, I. Uchida, A conductometric glucose sensor based on a twin-microband electrode coated with a polyaniline thin film. Bull. Chem. Soc. Japan **67**(10), 2864–2866 (1994)

15. S. Mansouri, J.S. Schultz, A miniature optical glucose sensor based on affinity binding. Nat. Biotechnol. **2**(10), 885–890 (1984)

16. N.D. Evans, D.J.S. Birch, O.J. Rolinski, J.C. Pickup, F. Hussain, Fluorescence-based glucose sensors. Biosens. Bioelectron. **20**(12), 2555–2565 (2004)

17. Y.-B. Hahn, R. Ahmad, N. Tripathy, Chemical and biological sensors based on metal oxide nanostructures. Chem. Commun. **48**, 10369–10385 (2012)

18. M.M. Rahman, A.J.S. Ahammad, J. Jin, S.J. Ahn, J.-J. Lee, A Comprehensive review of glucose biosensors based on nanostructured metal-oxides. Sensors **10**, 4855–4886 (2010). https://doi.org/10.3390/s100504855

19. C. Espro, N. Donato, S. Galvagno, D. Alosiso, Salvatore G. Leonardi, G. Neri, CuO nanowires-based electrodes for glucose sensors. Chem. Eng. Transact. **41**, 415–420 (2014)

20. C. Kong, L. Tang, X. Zhang, S. Sun, S. Yang, X. Song, Z. Yang, Templating synthesis of hollow CuO polyhedron and its application for one enzymatic glucose detection. J. Mater. Chem. A (2014). https://doi.org/10.1039/c4ta00703d

21. M.-J. Song, S.-K. Lee, J.-H. Kim, D.-S. Lim, Non-enzymatic glucose sensor based on Cu electrode modified with CuO nanoflowers. J. Electrochem. Soc. **160**, B43–B46 (2013)

22. N.M. Ahmad, J. Abdullah, N.I. Ramli, S. Abd Rahman, N.E. Azmi, Z. Hamzah, A. Saat, N.H. Rahman, Characterization of ZrO2/PEG composite film as immobilization matrix for glucose oxidase. World Academy of Science. Eng. Technol. Int. J. Mater. Metall. Eng. **7**, 8 (2013)

23. A.T.E. Viliana, S.-M. Chena, M.A. Ali, F.M.A. Al-Hemaid, Direct electrochemistry of glucose oxidase immobilized on ZrO$_2$ nanoparticles decorated reduced graphene oxide sheets for a glucose biosensor. RSC Adv. **4**, 30358–30367 (2014)

24. B. Wang, S. Li, J. Liu, M. Yu, Preparation of nickel nanoparticle/graphene composites for non-enzymatic electrochemical glucose biosensor applications. Mater. Res. Bull. **49**, 521–524 (2014)

25. Z. Yang, Y. Xu, J. Li, Z. Jian, S. Yu, Y. Zhang, X. Hu, D.D. Dionysiou, An enzymatic glucose biosensor based on a glassy carbon electrode modified with cylinder-shaped titanium dioxide nanorods. Microchim. Acta **182**(9–10), 1841–1848 (2015)

26. N. Haghighi, R. Hallaj, A. Salimi, Immobilization of glucose oxidase onto a novel platform based on modified TiO$_2$ and graphene oxide, direct electrochemistry, catalytic and photocatalytic activity. Mater. Sci. Eng., C **73**, 417–424 (2017)

27. S. Saha, S.K. Arya, S.P. Singh, B.D. Malhotra, K. Sreenivas, V. Gupta, Cerium oxide (CeO$_2$) thin film for mediator-less glucose biosensors, in *Materials Research Society Symposium Proceedings* (2009)

28. D. Patil, N.Q. Dung, H. Jung, S.Y. Ahn, D.M. Jang, D. Kim, Enzymatic glucose biosensor based on CeO$_2$ nanorods synthesized by non-isothermal precipitation. Biosens. Bioelectr. **31**(1), 176–181 (2012)

29. A. Kaushik, R. Khan, P.R. Solanki, P. Pandey, J. Alam, S. Ahmad, B.D. Malhotra, Iron oxide nanoparticles–chitosan composite based glucose biosensor. Biosens. Bioelectron. **24**(4), 676–683 (2008)
30. Ç. Atan, E. Karaku, Novel zinc oxide nanorod and chitosan-based electrochemical glucose biosensors for glucose assay in human serum samples. Sens. Lett. **12**(11), 1613–1619 (2014)
31. Q. Ma, K. Nakazato, Low-temperature fabrication of ZnO nanorods/ferrocenyl–alkanethiol bilayer electrode and its application for enzymatic glucose detection. Biosens. Bioelectr. **51**, 362–365 (2014)
32. M.I. Said, H. Azza, R. Fatma, A.M. Abdel-aal, Fabrication of novel electrochemical sensors based on modification with different polymorphs of MnO_2 nanoparticles. RSC Adv. **8**, 18698–18713 (2018)
33. E.O. Fayemi, A.S. Adekunle, E.E. Ebenso, Metal oxide nanoparticles/multi-walled carbon nanotube nanocomposite modified electrode for the detection of dopamine: comparative electrochemical study. J. Biosens. Bioelectr. **6**, 190 (2015). https://doi.org/10.4172/2155-6210.1000190
34. M.J. Devine, H. Plun-Favreau, N.W. Wood, Parkinson's disease and cancer: two wars, one front. Nat. Rev. Cancer **11**(11), 812–823 (2011)
35. S. Shahrokhian, E. Asadian, Electrochemical determination of l-dopa in the presence of ascorbic acid on the surface of the glassy carbon electrode modified by a bilayer of multi-walled carbon nanotube and poly-pyrrole doped with tiron. J. Electroanal. Chem. **636**(1–2), 40–46 (2009)
36. H. Beitollahi, F. Garkani, *Graphene oxide/ZnO Nano Composite for Sensitive and Selective Electrochemical Sensing of Levodopa and Tyrosine Using Modified Graphite Screen Printed Electrode* (2016), pp. 1–9
37. L. P. Martin, R. S. Glass, *Hydrogen Sensor Based on YSZ Electrolyte and Tin-Doped Indium Oxide Electrode* (2015), pp. 43–47
38. G. Lu, N. Miura, N. Yamazoe, High-temperature hydrogen sensor based on stabilized zirconia and a metal oxide electrode. Sens. Actuators B: Chem. **36**, 130–135 (1996)
39. S. Ayu, M. Breedon, N. Miura, Sensing characteristics of aged zirconia-based hydrogen sensor utilizing Zn–Ta-based oxide sensing-electrode. Electrochem. Commun. **31**, 133–136 (2013)
40. J. Yi, H. Zhang, Z. Zhang, D. Chen, Hierarchical porous hollow SnO_2 nanofiber sensing electrode for high performance potentiometric H_2 sensor. Sens. Actuators B Chem. **268**, 456–464 (2018)
41. Y. Li, X. Li, Z. Tang, J. Wang, J. Yu, Z. Tang, Potentiometric hydrogen sensors based on yttria-stabilized zirconia electrolyte (YSZ) and $CdWO_4$ interface. Sens. Actuators B. Chem. **223**, 365–371 (2016)
42. S.A. Anggraini, M. Breedon, N. Miura, Effect of sintering temperature on hydrogen sensing characteristics of zirconia sensor utilizing Zn–Ta–O-based sensing electrode. J. Electrochem. Soc. **160**(9), B164–B169 (2013)
43. J. Yu, J. Yang, Z. Tang, Z. Tang, J. Wang, X. Li, Mixed potential hydrogen sensor using $ZnWO_4$ sensing electrode. Sens. Actuators B Chem. **195**, 520–525 (2014)
44. H. Zhang, J. Yi, X. Jiang, Fast response, highly sensitive and selective mixed-potential H2 sensor based on $(La, Sr)(Cr, Fe)O_3$-δ perovskite sensing electrode. ACS Appl. Mater. Interfaces. 3–10 (2017)
45. Y. Li, X. Li, Z. Tang, Z. Tang, J. Yu, J. Wang, Hydrogen sensing of the mixed-potential-type $MnWO_4$/YSZ/Pt sensor. Sens. Actuators, B Chem. **206**, 176–180 (2015)
46. N. Miura, T. Sato, S.A. Anggraini, A review of mixed-potential type zirconia-based gas sensors. Ionics **20**, 901–925 (2014)
47. J.W. Yoon, M.L. Grilli, E. Di Bartolomeo, R. Polini, E. Traversa, The NO_2 response of solid electrolyte sensors made using nano-sized $LaFeO_3$ electrodes. Sens. Actuators B: Chem. **76**(2), 483–488 (2001)
48. W. Xiong, G.M. Kale, Electrochemical NO_2 sensor using a $NiFe_{1.9}Al_{0.1}O_4$ oxide spinel electrode. Anal. Chem. **79**(10), 3561–3567 (2007)

49. N. Miura, J. Wang, M. Nakatou, P. Elumalai, S. Zhuiykov, M. Hasei, High-temperature operating characteristics of mixed-potential-type NO_2 sensor based on stabilized-zirconia tube and NiO sensing electrode. Sens. Actuators B Chem. **114**(2), 903–909 (2006)
50. A. Morata, J.P. Viricelle, A. Taranc, Development and characterisation of a screen-printed mixed potential gas sensor. Sens. Actuators B: Chem. **130**, 561–566 (2008)
51. Y. Fujio, V.V. Plashnitsa, M. Breedon, N. Miura, Construction of sensitive and selective zirconia-based CO sensors using $ZnCr_2O_4$-based sensing electrodes. Langmuir **28**(2), 1638–1645 (2012)
52. B. Sljuki, C.E. Banks, A. Crossley, R.G. Compton, Lead (IV) oxide—graphite composite electrodes: application to sensing of ammonia, nitrite and phenols. Analytica Chimica Acta **587**, 240–246 (2007)
53. K. Singh, A.A. Ibrahim, A. Umar, A. Kumar, G.R. Chaudhary, S. Singh, S.K. Mehta, Synthesis of CeO_2–ZnO nanoellipsoids as potential scaffold for the efficient detection of 4-nitrophenol. Sens. Actuators B Chem. **202**, 1044–1050 (2014)
54. Z. Liu, J. Du, C. Qiu, L. Huang, H. Ma, D. Shen, Y. Ding, Electrochemical sensor for detection of p-nitrophenol based on nanoporous gold. Electrochem. Commun. **11**(7), 1365–1368 (2009)
55. N. Lezi, A. Economou, J. Barek, M. Prodromidis, Screen-printed disposable sensors modified with bismuth precursors for rapid voltammetric determination of 3 ecotoxic nitrophenols. Electroanalysis **26**, 766–775 (2014)
56. M.M. Rahman, S.B. Khan, A.M. Asiri, A.G. Al-Sehemi, Chemical sensor development based on polycrystalline gold electrode embedded low-dimensional Ag_2O nanoparticles. Electrochim. Acta **112**, 422–430 (2013)
57. J. Wu, Q. Wang, A. Umar, S. Sun, L. Huang, J. Wang, Y. Gao, Highly sensitive p-nitrophenol chemical sensor based on crystalline α-MnO_2 nanotubes. New J. Chem. **38**(9), 4420–4426 (2014)
58. M.M. Rahman, G. Gruner, M.S. Al-Ghamdi, M.A. Daous, S.B. Khan, A.M. Asiri, Chemosensors development based on low-dimensional codoped Mn_2O_3–ZnO nanoparticles using flat-silver electrodes. Chem. Cent. J. **7**(1), 60 (2013)
59. M. Abaker G.N. Dar, A.A. Umar, S.A. Zaidi, A.A. Ibrahim, S. Baskoutas, A. Al-Hajry, CuO nanocubes based highly-sensitive 4-nitrophenol chemical sensor. Sci. Adv. Mater. **4**(8), 893–900 (2012)
60. Y. Haldorai, K. Giribabu, S. Hwang, C.H. Kwak, Y.S. Huh, Y.-K. Han, Facile synthesis of α-MnO_2 nanorod/graphene nanocomposite paper electrodes using a 3D precursor for supercapacitors and sensing platform to detect 4-nitrophenol. Electrochim. Acta **222**, 717–727 (2016)
61. T. Kooyers, W. Westerhof, Toxicology and health risks of hydroquinone in skin lightening formulations. J. Eur. Acad. Dermatol. Venereol. (2005)
62. M.U.A. Prathap, B. Satpati, R. Srivastava, Facile preparation of polyaniline/MnO_2 nanofibers and its electrochemical application in the simultaneous determination of catechol, hydroquinone, and resorcinol. Sens. Actuators B Chem. **186**, 67–77 (2013)
63. T. Gan, J. Sun, K. Huang, L. Song, Y. Li, A graphene oxide–mesoporous MnO_2 nanocomposite modified glassy carbon electrode as a novel and efficient voltammetric sensor for simultaneous determination of hydroquinone and catechol. Sens. Actuators B Chem. **177**, 412–418 (2013)
64. A. Umar, A. Al-Hajry, R. Ahmad, S.G. Ansari, M.S. Al-Assiri, H. Algarni, Fabrication and characterization of a highly sensitive hydroquinone chemical sensor based on iron-doped ZnO nanorods. Dalt. Trans. **44**(48), 21081–21087 (2015)
65. S. Ameen, M.S. Akhtar, H. Shik, Highly dense ZnO nanowhiskers for the low level detection of p-hydroquinone. Mater. Lett. 1–5 (2015)
66. B. Unnikrishnan, P. Ru, S. Chen, Electrochemically synthesized Pt–MnO_2 composite particles for simultaneous determination of catechol and hydroquinone. Sens. Actuators B. Chem. **169**, 235–242 (2012)
67. L. Yang, H. Zhao, S. Fan, B. Li, C. Li, A highly sensitive electrochemical sensor for simultaneous determination of hydroquinone and bisphenol A based on the ultrafine Pd nanoparticle@TiO_2 functionalized SiC. Anal. Chim. Acta **852**, 28–36 (2014)

68. S. Erogul, S.Z. Bas, M. Ozmen, S. Yildiz, A new electrochemical sensor based on Fe_3O_4 functionalized graphene oxide-gold nanoparticle composite film for simultaneous determination of catechol and hydroquinone. Electrochim. Acta **186**, 302–313 (2015)
69. S. Hilliard, G. Baldinozzi, D. Friedrich, S. Kressman, H. Strub, V. Artero, C. Laberty-Robert, Mesoporous thin film WO_3 photoanode for photoelectrochemical water splitting: a sol–gel dip coating approach. Sustain. Energy Fuels **1**, 145–153 (2017)
70. M. Kitano, K. Tsujimaru, M. Anpo, Hydrogen production using highly active titanium oxide-based photocatalysts. Top. Catal. **49**(1–2), 4–17 (2008)
71. A. Kudo, Y. Miseki, Heterogeneous photocatalyst materials for water splitting. Chem. Soc. Rev. **38**, 253–278 (2009)
72. Y. Wang, C. Sun, X. Zhao, B. Cui, Z. Zeng, A. Wang, G. Liu, H. Cui, The application of nano-TiO_2 photo semiconductors in agriculture. Nanoscale Res. Lett. **11**(1), 529 (2016)
73. H.N. Guan, D.F. Chi, J. Yu, X.C. Li, A novel photodegradable insecticide: preparation, characterization and properties evaluation of nano-Imidacloprid. P-B. Physiol. **92**(2), 83–91 (2008)
74. D.C. Lim, J.H. Jeong, K. Hong, S. Nho, J.-Y. Lee, Q.V. Hoang, S.K. Lee, K. Pyo, D. Lee, S. Cho, Semi-transparent plastic solar cell based on oxide-metal-oxide multilayer electrodes. Prog. Photovoltaics Res. Appl. **26**(3), 188–195 (2018)
75. M. Zhi, C. Xiang, J. Li, M. Li, N. Wu, Nanostructured carbon-metal oxide composite electrodes for supercapacitors: a review. Nanoscale **5**(1), 72–88 (2013)
76. V.D. Patake, C.D. Lokhande, O.S. Joo, Electrodeposited ruthenium oxide thin films for supercapacitor: effect of surface treatments. Appl. Surf. Sci. **255**(7), 4192–4196 (2009)
77. D. Yan, Z. Guo, G. Zhu, Z. Yu, H. Xu, A. Yu, MnO_2 film with three-dimensional structure prepared by hydrothermal process for supercapacitor. J. Power Sources **199**, 409–412 (2012)
78. U.M. Patil, R.R. Salunkhe, K.V. Gurav, C.D. Lokhande, Chemically deposited nanocrystalline NiO thin films for supercapacitor application. Appl. Surf. Sci. **255**(5, part 2), 2603–2607 (2008)
79. X.C. Dong, H. Xu, X.-W. Wang, Y.-X. Huang, M.B. Chan-Park, H. Zhang, L.-H. Wang, W. Huang, P. Chen, 3D graphene-cobalt oxide electrode for high-performance supercapacitor and enzymeless glucose detection. ACS Nano **6**(4), 3206–3213 (2012)
80. J. Mu, B. Chen, Z. Guo, M. Zhang, Z. Zhang, C. Shao, Y. Liu, Tin oxide (SnO_2) nanoparticles/electrospun carbon nanofibers (CNFs) heterostructures: Controlled fabrication and high capacitive behavior. J. Colloid Interface Sci. **356**(2), 706–712 (2011)
81. X. Zhang, W. Shi, J. Zhu, D.J. Kharistal, W. Zhao, B.S. Lalia, H.H. Hng, Q. Yan, High-power and high-energy-density flexible pseudocapacitor electrodes made from porous CuO nanobelts and single-walled carbon nanotubes. ACS Nano **5**(3), 2013–2019 (2011)
82. C. Xiang, M. Li, M. Zhi, A. Manivannan, N. Wu, Reduced graphene oxide/titanium dioxide composites for supercapacitor electrodes: shape and coupling effects. J. Mater. Chem. **22**(36), 19161 (2012)
83. X. Dong, Y. Cao, J. Wang, M.B. Chan-Park, L. Wang, W. Huang, P. Chen, Hybrid structure of zinc oxide nanorods and three dimensional graphene foam for supercapacitor and electrochemical sensor applications. RSC Adv. **2**(10), 4364 (2012)
84. X. Lu, T. Zhai, X. Zhang, Y. Shen, L. Yuan, B. Hu, L. Gong, J. Chen, Y. Gao, J. Zhou, Y. Tong, Z.L. Wang, WO_3-x@Au@MnO_2 core-shell nanowires on carbon fabric for high-performance flexible supercapacitors. Adv. Mater. **24**(7), 938–944 (2012)
85. X. Liu, N. Zhang, J. Ni, L. Gao, Improved electrochemical performance of sol-gel method prepared $Na_4Mn_9O_{18}$ in aqueous hybrid Na-ion supercapacitor. J. Solid State Electrochem. **17**(7), 1939–1944 (2013)
86. P. Chen, G. Shen, Y. Shi, H. Chen, C. Zhou, Preparation and characterization of flexible asymmetric supercapacitors. ACS Nano **4**(8), 4403–4411 (2010)
87. L.J. Hannah, *Climate Change Biology* (Academic Press, Cambridge, 2010). https://doi.org/10.1016/C2013-0-12835. ISBN 978-0-12-420218-4
88. J. Kasnatscheew, M. Evertz, B. Streipert, R. Wagner, S. Nowak, L.I. Cekic, M. Winter, Improving cycle life of layered lithium transition metal oxide ($LiMO_2$) based positive electrodes for

Li ion batteries by smart selection of the electrochemical charge conditions. J. Power Sources **359**, 458–467 (2017)

89. J. Mao, J. Iocozzia, J. Huang, K. Meng, Y. Lai, Z. Lin, Graphene aerogels for efficient energy storage and conversion. E. & E. Sci. **11**(4), 772–799 (2018)
90. D. Cericola, P. Ruch, R. Kötz, P. Novák, A. Wokaun, Simulation of a supercapacitor/Li-ion battery hybrid for pulsed applications. J. Power Sources **195**(9), 2731–2736 (2010)
91. M. Stoller, R. Ruoff, Best practice methods for determining an electrode material's performance for ultracapacitors. E. & E. Sci. **3**(9), 1294 (2010)
92. H. Jiang, J. Ma, C. Li, Mesoporous carbon incorporated metal oxide nanomaterials as supercapacitor electrodes. Adv. Mater. **24**(30), 4197–4202 (2012)
93. S.P.S. Badwal, S.S. Giddey, C. Munnings, A.I. Bhatt, A.F. Hollenkamp, Emerging electrochemical energy conversion and storage technologies. Front. Chem. **2** (2014)
94. A. Vlad, N. Singh, C. Galande, P.M. Ajayan, Design considerations for unconventional electrochemical energy storage architectures. Adv. Energy Mater. **5**(19), 1402115 (2015)
95. H. Zhang, H. Zhao, M. Khan, W. Zou, J. Xu, L. Zhang, J. Zhang, Recent progress in advanced electrode materials, separators and electrolytes for lithium batteries. J. Mater. Chem. A, **6**(42), 20564–20620 (2018)
96. Z. Wang, L. Zhou, X.W. David Lou, Metal oxide hollow nanostructures for lithium-ion batteries. Adv. Mater. **24**(14), 1903–1911 (2012)
97. X. Huang, Z. Yin, S. Wu, X. Qi, Q. He, Q. Zhang, Q. Yan, F. Boey, H. Zhang, Graphene-based materials: synthesis, characterization, properties, and applications. Small **7**(14), 1876–1902 (2011)
98. J. Jiang, Y. Li, J. Liu, X. Huang, C. Yuan, X.W.D. Lou, Recent advances in metal oxide-based electrode architecture design for electrochemical energy storage. Adv. Mater. **24**(38), 5166–5180 (2012)
99. M. Notarianni, J. Liu, K. Vernon, N. Motta, Synthesis and applications of carbon nanomaterials for energy generation and storage. Beilstein J. Nanotechnol. **7**, 149–96 (2016)
100. B. Li, X. Shao, Y. Hao, Y. Zhao, Ultrasonic-spray-assisted synthesis of metal oxide hollow/mesoporous microspheres for catalytic CO oxidation. RSC Adv. **5**(104), 85640–85645 (2015)
101. F. Cheng, J. Chen, Transition metal vanadium oxides and vanadate materials for lithium batteries. J. Mater. Chem. **21**(27), 9841 (2011)
102. X. Xia, Y. Zhang, D. Chao, C. Guan, Y. Zhang, L. Li, et al., Solution synthesis of metal oxides for electrochemical energy storage applications. Nanoscale **6**(10), 5008–5048 (2014)
103. S. Yan, K. Abhilash, L. Tang, M. Yang, Y. Ma, Q. Xia, Q. Guo, H. Xia, Research advances of amorphous metal oxides in electrochemical energy storage and conversion. Small 1804371 (2018)
104. C.M. Julien, A. Mauger, Nanostructured MnO_2 as electrode materials for energy storage. Nanomaterials **7**(11), 396 (2017)
105. B. Scrosati, J. Garche, Lithium batteries: status, prospects and future. J. Power Sources **195**(9), 2419–2430 (2010)
106. P. Polzot, S. Laruelle, S. Grugeon, L. Dupont, J. Tarascon, Nano-sized transition-metal oxides as negative-electrode materials for lithium-ion batterles. Nat. Publ. Gr. **407**(6803), 496–499 (2000)
107. J. Chen, L. Archer, X.L. Wen (David), SnO_2 hollow structures and TiO_2 nanosheets for lithium-ion batteries. J. M. Chem. **21**(27), 9912 (2011)
108. Y. Li, J. Shi, Hollow-structured mesoporous materials: chemical synthesis, functionalization and applications. Adv. Mater. **26**(20), 3176–3205 (2014)
109. B. Zhao, L.C. Lee, L. Yang, A.J. Pearson, H. Lu, X.J. She, L. Cui, K.H.L. Zhang, R.L.Z. Hoye, A. Karani, P. Xu, A. Sadhanala, N.C. Greenham, R.H. Friend, J.L. MacManus-Driscoll, D. Di, In situ atmospheric deposition of ultrasmooth nickel oxide for efficient perovskite solar cells. ACS Appl. Mater. Interfaces **10**(49), 41849–41854 (2018)
110. W. Chen, Y. Qiu, Y. Zhong, K.S. Wong, S. Yang, High-efficiency dye-sensitized solar cells based on the composite photoanodes of SnO_2 nanoparticles/ZnO nanotetrapods. J. Phys. Chem. A **114**(9), 3127–3138 (2010)

111. W. Chen, Y. Qiua, S. Yang, A new ZnO nanotetrapods/SnO$_2$ nanoparticles composite photoanode for high efficiency flexible dye-sensitized solar cells. Phys. Chem. Chem. Phys. **12**, 9494–9501 (2010)
112. Y. Chiba, A. Islam, Y. Watanabe, R. Komiya, N. Koide, L. Han, Dye-sensitized solar cells with conversion efficiency of 11.1%. Jpn. J. Appl. Phys. **45**(25), L638–L640 (2006)
113. T. Hu, T. Becker, N. Pourdavoud, J. Zhao, K.O. Brinkmann, R. Heiderhoff, T. Gahlmann, Z. Huang, S. Olthof, K. Meerholz, D. Többens, B. Cheng, Y. Chen, T. Riedl, Indium-free perovskite solar cells enabled by impermeable tin-oxide electron extraction layers. Adv. Mater. **29**(27), 1606656 (2017)
114. X. Zhang, C. Hägglund, M.B. Johansson, K. Sveinbjörnsson, E.M.J. Johansson, Fine tuned nanolayered metal/metal oxide electrode for semitransparent colloidal quantum dot solar cells. Adv. Funct. Mater. **26**(12), 1921–1929 (2016)
115. K. Zilberberg, F. Gasse, R. Pagui, A. Polywka, A. Behrendt, S. Trost, R. Heiderhoff, P. Görrn, T. Riedl, Highly robust indium-free transparent conductive electrodes based on composites of silver nanowires and conductive metal oxides. Adv. Funct. Mater. **24**(12), 1671–1678 (2014)

Chapter 9
Application of Modified Metal Oxide Electrodes in Photoelectrochemical Removal of Organic Pollutants from Wastewater

William Wilson Anku, Onoyivwe Monday Ama, Suprakas Sinha Ray, and Peter Ogbemudia Osifo

Abstract The scarcity of clean water due to the increase in ground and surface water pollution by numerous pollutants from municipal, industrial, and agricultural sources is considered to be the most pressing environmental problem and a threat to the survival of humans. Among the methods being tested for the sustainable removal of these pollutants from water before use or disposed into the environment is photoelectrochemical degradation. This technique is a combination of electrochemical oxidation and heterogeneous photocatalytic degradation and involves the use of metal oxide semiconductor-based electrodes in the presence of light. This process begins with the generation of electrons by the metal oxide semiconductors which react with oxygen and water molecules to produce oxidants including superoxide and hydroxyl radicals which are responsible for the degradation of the pollutants. The efficacy of this process is, however, hampered by the high rate at which the electrons recombine with the co-generated holes and the poor visible light activity of the semiconductors due to their wide band gaps. A number of modifications have been made to the semiconductors to resolve the above-mentioned problems and improve on the efficiency of the photoelectrochemical degradation process. This chapter dwells mainly on the various modifications that have been made to the metal oxides including the use of carbon-based and polymeric materials, doping with metal, doping with non-metals, co-doping with metals and non-metals and formation of mixed metal oxide/heterostructures.

W. W. Anku (✉)
CSIR-Water Research Institute, P.O. Box M.32, Accra, Ghana
e-mail: williamanku85@gmail.com

O. M. Ama · S. S. Ray
Department of Chemical Science, University of Johannesburg, Doornfontein, 2028, Johannesburg, South Africa

DST-CSIR National Center for Nanostructured Materials, Council for Scientific and Industrial Research, Pretoria 0001, South Africa

O. M. Ama · P. O. Osifo
Department of Chemical Engineering, Vaal University of Technology, Private Mail Bag X021, Vanderbijlpark 1900, South Africa

© Springer Nature Switzerland AG 2020
O. M. Ama and S. S. Ray (eds.), *Nanostructured Metal-Oxide Electrode Materials for Water Purification*, Engineering Materials,
https://doi.org/10.1007/978-3-030-43346-8_9

9.1 Introduction

The difficulty in providing adequate amount of clean water to meet the demands of individual homes and industries is a known problem across the globe. This difficulty is mainly due to the pollution of water bodies with organic and inorganic pollutants through the release of raw industrial, domestic, agricultural, hospital and pharmaceutical industrial wastewater into the environment [1]. Utilization of water bodies containing pollutants such as dyes, heavy metals, pesticides and so on has detrimental effects on human health and aquatic organisms [1]. Thus, it has become necessary to treat polluted water before being used for domestic and industrial purposes and/or before it is finally released into receiving water bodies.

Conventional methods such as adsorption, coagulation and flocculation, precipitation, activated sludge digestion, anaerobic digestion and so on designed for treating water have been identified to be ineffective especially with regards to the removal of emerging organic pollutants [2]. This necessitated the development of alternative water treatment technologies to complement the existing conventional methods.

One of the alternative water treatment methods with known effectiveness is advanced oxidation processes [3]. The main feature of these processes is the production of hydroxyl radicals ($\cdot OH$) that oxidise the toxic pollutants into harmless ones [3]. Among the advanced oxidation processes is electrochemical oxidation, where the hydroxyl radicals required for the degradation of the pollutants are generated electronically at an anode. Although electrochemical oxidation is considered to have many advantages over other advanced oxidation processes including ease of automation, cost minimisation, and chemical stability [4], its efficiency is hampered by some drawbacks. These drawbacks include the fact that the process involves a mass transfer of the pollutants from the solution onto the surface of the electrode. Another problem is the evolution of oxygen during the degradation process due to the application of high voltage. This oxygen evolution process occurs alongside, and interferes with hydroxyl radical production and hence retards the pollutant degradation process [5].

An appropriate approach that has been adopted to overcome the drawbacks of electrochemical oxidation process is by combining it with photocatalytic degradation through the use of metal oxide semiconductors. In the combined electrochemical and photocatalytic degradation process (preferably referred to as photoelectrochemical degradation) electrons are promoted from the valence band to the conduction band of the metal oxide semiconductors upon light exposure, leaving holes (positive charges) at the valence bands. The generated electrons react with the oxygen molecules to form superoxide radicals ($\cdot O_2{}^-$), thereby preventing the oxygen molecules from interfering with hydroxyl radical production in the electrochemical process [6]. In addition, the holes react with water molecules to produce more hydroxyl radicals ($\cdot OH$). The $\cdot O_2{}^-$ and $\cdot OH$ which are excellent oxidants then oxidise the organic pollutants into harmless carbon dioxide and water.

The efficiencies of the metal oxide semiconductors used in this process are, however, limited by their slow electrons transfer rate, fast recombination of the generated electrons with the holes, and limited visible light absorption due to their wide band gaps [7, 8]. In order to solve these problems, several modifications have been made to the traditional metal oxide semiconductor photocatalysts. This chapter dwells more on the various modifications with details presented in the subsequent sections.

9.2 Methods of Modification of Metal Oxides for Improved Photoelectrochemical Activity

Major research activities have been undertaken to develop metal oxides with reduced recombination rate of photogenerated electrons and holes, reduced band gap and the subsequent improvement in their visible light responsivity. Positive results have been attained in this regard through the utilisation of a variant of techniques intended to modify the optical and electronic properties of metal oxides for efficient photoelectrochemical degradation process. These techniques comprise doping with metals [9, 10], doping with non-metal [11, 12], sensitization with dyes [13], use of mixed metal oxides (metal oxides coupling) [14, 15], metal and non-metal co-doping [16] and combination with carbon based materials. Detailed discussions of these techniques are presented below.

9.2.1 Modification with Carbon-Based Materials

9.2.1.1 Modification with Graphene Materials

Graphene is an allotrope of carbon with sp^2 hybridized carbon atoms that are organised into a 2D honeycomb lattice with a C–C bond length of 0.142 nm (Fig. 9.1) [16].

Fig. 9.1 Structure of graphene

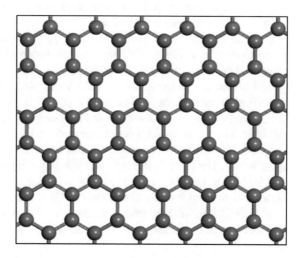

It is a zero-gap semiconductor with a conical band structure and excellent in-plane conductivity [17]. Graphene sheets possess some peculiar properties that make them applicable in electrodes for photoelectrochemical degradation. Among these properties are large surface area, greater elasticity, high mechanical strength, and high electron mobility at room temperature [18]. In addition, graphene has the tendency to display voltage differences in the presence of a magnetic field at room temperature. It also exhibits excellent optical properties and tunable band gap with no existence of electron states [18]. The comparatively low cost of production coupled with its environmentally friendly nature makes it more appropriate and desirable material for the fabrication of electrodes [18]. Due to its zero band gap which makes it behave like a metal, and its hydrophobicity, graphene is preferably converted into other graphene-based materials including graphene oxide and reduced graphene oxide prior to its application in various processes including photoelectrochemical degradation.

Modification with Reduced Graphene Oxide

Graphene oxide (GO) is one of the graphene-related materials produced through oxidation/functionalisation of graphene. Due to the existence of both hydrophobic and hydrophilic functional groups in its structure, GO disperses easily in water, making it compatible with biological molecules couple with the ability to bind with a wide range of other molecules. However, GO possesses poor electrical conductivity, making its use in photoelectrochemical process a daunting task. In order to improve on its electrical conductivity, GO is converted to reduced graphene oxide through chemical reduction or deoxygenation.

Reduced graphene oxide (rGO), an example of graphene-related materials, is viewed as the more suitable form of graphene materials for electrochemical application owing to its high electrical conductivity, and the fact that the deoxygenation procedure enhances immobilization of other materials/elements onto its surface [19]. As a result, a number of rGO based electrodes have been fabricated and applied in the removal of pollutants from water.

A study conducted by Umukoro et al. [20] involved the immobilization of silver-doped zinc oxide on rGO and the conversion of the composite into a photoanode. The fabricated photoanode was subsequently employed for the degradation of orange II dye in water. The rGO based photoanode was noted to demonstrate impressive performance with 93% orange II dye removal efficiency. Similarly, Moraes et al. [21] described the fabrication of a ruthenium oxide-reduced graphene oxide (RuO_2–rGO) based electrode and its application in photoelectrochemical degradation of 17 B-estradiol in water under light irradiation with encouraging results. The analysis showed that the RuO_2–rGO electrode was more effective than the bare RuO_2 electrode as the RuO_2–rGO electrode displayed a comparatively higher photoelectrochemical activity with 92% efficiency in 60 min.

In a similar manner, methanol, ethanol, methyl orange, phenol and polyvinyl alcohol removal from water was performed through the use of an rGO–CdS–H_2W_{12}

composite film electrode under irradiation [22]. According to their result, rGO introduction improved the separation and transportation of the charges, causing the photocurrent response of the CdS–H_2W_{12} to improved significantly after its modification with rGO [22]. The fabrication and application of an FeOOH functionalized phosphorus doped bismuth vanadate coated rGO (FeOOH/P:BiVO$_4$/rGO) composite electrode for the degradation of 2,4-dichlorophenol has been carried out by Shi et al. [23]. The FeOOH/P:BiVO$_4$/rGO exhibited excellent photoelectrochemical performance against the pollutant as a result of improved suppression of electron-hole recombination rate and efficient charge transfer at the metal oxide-electrolyte interface.

Modification with Exfoliated Graphite

Exfoliated graphite (EG) is another form of graphene-based material obtained through intercalation of natural graphite using acid mixtures such as H_2SO_4–HNO_3 mixture (3:1 v/v) and subsequent heating at high temperature (around 800 °C) for about 1 min. Because of its impressive electrocatalytic activity, large surface area for high pollutants adsorption, and better homogeneity coupled with the fact that it can be processed into electrodes without the need for a binder [24], EG is considered appropriate material for application in photoelectrochemical degradation processes.

In order to assess the photoelectrocatalytic ability of an EG based electrode, Ntsendwana et al. [25] fabricated an EG-diamond electrode and used it in the degradation of trichloroethylene in the presence of Na_2SO_4 electrolyte. Comparatively, the EG–diamond, displayed higher trichloroethylene removal efficiency that the pure EG. The EG-diamond removed 94% while the pure EG removed 57% of the pollutant in water. They considered the use of the EG-diamond electrode pollutants removal process as a cost-effective approach due to the ease with which the composite electrode can be prepared. EG-WO$_3$ composite electrode meant for the removal of eosin yellow dyes from water has also been developed [26]. The pollutant removal study was performed through photolysis, photocatalysis, and photoelectrochemical degradation methods and the results were compared. The analysis identified photoelectrochemical degradation as the most appropriate technique for eosin yellow removal from polluted water.

In a similar vein, Ama et al. [27] proposed a photoanode consisting of EG, copper oxide, and zinc oxide (EG-CuO/ZnO) for use in the degradation of 4-nitrophenol from simulated wastewater. Individual EG electrode, CuO/ZnO electrode, and EG-CuO/ZnO electrodes were used in a comparative degradation of the dye with the EG-CuO/ZnO electrode exhibiting the highest efficiency. In addition, the most effective electrode (EG-CuO/ZnO) was in turn utilised in electrochemical and photoelectrochemical degradation of the dye and the results were compared. The photoelectrochemical technique displayed higher efficiency when compared with the electrochemical method. It was identified that modifying the CuO–ZnO composite with EG endowed the composite with large surface area and efficient visible light absorptivity. A photoanode consisting of EG and Ti-doped RuO_2 has also been prepared and applied in the degradation of phenol with impressive result [28].

9.2.1.2 Modification with Carbon Nanotubes

Modification of metal oxide semiconductors with carbon nanotubes (CNTs) is recognised an important approach for enhancing the electrochemical performance of the metal oxides. Because, CNTs are endowed with large surface area, outstanding electrical conductivity, excellent chemical and thermal stability coupled with good mechanical characteristics, their combination with metal oxides results in a photoanode with improved visible and ultraviolet lights sensitivity [29, 30]. This excellent light utilisation leads to profound photoelectrochemical performances. In composites of this nature, the CNT performs the role of a photosensitizer, pollutant adsorbent, as well as a trap site for the generated electrons. Additionally, the large surface area of the CNT permits even distribution of the nanomaterials on its surface, promoting efficient pollutant adsorption and degradation [31]. Effective interaction between the metal oxide nanoparticles and the CNT ensures a proficient transfer of electrons leading to the reduction in the recombination rate of the electrons and holes [32] and the resultant improvement in photoelectrochemical activity of the electrode.

A photoanode based on multiwalled carbon nanotube modified bismuth vanadate ($BiVO_4$/MWCNT) was synthesised by Ribeiro et al. [33] and applied for the removal of 4-nitrophenol from water under visible light irradiation. The $BiVO_4$/MWCNT based electrode displayed higher photocurrent compared to the bare $BiVO_4$, an indication that the presence of the MWCNT led to an improvement in charge transfer and minimisation of the electrons and holes recombination rate in the electrode. They concluded that the synergistic properties of both $BiVO_4$ and MWCNT caused the enhanced 4-nitrophenol degradation performance of the resultant electrode. In addition, Chaudhary et al. [34] found a significant enhancement in the photoelectrochemical effectiveness of ZnO under visible light irradiation after incorporation of MWCNTs. Incorporation of MWCNTs with ZnO caused trapping of the photogenerated electrons resulting in their delayed recombination and the ensuing visible light boosted photoelectrocatalytic methylene blue degradation activity of the sample. In his work, Ivana [35] synthesised a TiO_2 coated MWCNTs and noted that all the MWCNTs modified electrodes exhibited higher photoelectrocatalytic activities compared to the bare TiO_2 when applied in the degradation of carbaryl pesticide upon exposure to simulated visible light. The excellent performance of the anode was attributed to the unique properties of its components. Supporting the TiO_2 on MWCNTs improved charge separation and minimised charge carrier recombination which would otherwise lower the effectiveness of the degradation process [35].

Jakab et al. [36] also studied the photoelectrochemical activity of MWCNTs and TiO_2 based electrode and showed that combination of MWCNTs with TiO_2 photocatalysts increased the visible light absorption spectrum of the resultant composite, and the TiO_2-MWCNTs composite demonstrated better electrochemical activity and degraded pentachlorophenol better than the pure TiO_2. In a similar manner, the photoelectrochemical degradation ability of a $CNT\text{-}TiO_2$ based photoanode against methylene blue dye was studied by Zhang et al. [37]. The study involved application of the $CNT\text{-}TiO_2$ composite in a comparative photocatalytic and photoelectrocatalytic oxidation of the pollutant with the photoelectrocatalytic procedure exhibiting the highest efficiency.

9.2.2   Formation of Mixed Metal Oxides/Heterostructures

Among the efforts being made to modify metal oxide semiconductors in order to enhance their properties is the coupling of two or more metal oxides possessing different band gaps (formation of heterostructures). This modification promotes efficient separation of charges and retards the rate at which they recombine leading to augmented electrochemical activity of the resultant heterostructure. The mixed metal oxides formation can be achieved in one of two ways: as coupled and capped mixed metal oxides [38].

With regards to the coupled system (Fig. 9.2), the metal oxides are in direct contact with one another. The holes and the electrons, upon generation, become available on the surfaces of all the metal oxides, permitting oxidation and reduction reactions on all of them. In the capped system of formation of mixed metal oxides (Fig. 9.3) the metal oxides are arranged in a core/shell form. The generated electrons get excited from the shell metal oxide to the energy level of the core and become trapped in

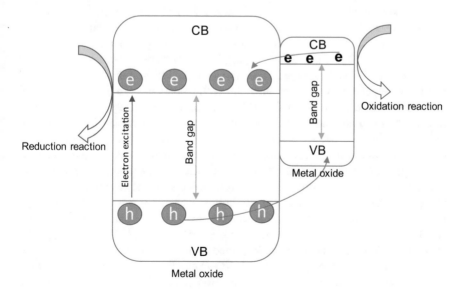

Fig. 9.2 Mechanism of charge transfer in a coupled semiconductor system

Fig. 9.3 Mechanism of
charge transfer in capped
semiconductor system

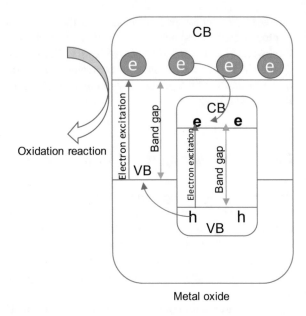

the core semiconductor system, thus rendering them inaccessible. Only the holes become available leading to selective oxidation process [39]. The unreachability of the generated electrons in this type of metal oxides arrangement renders it unsuitable for reactions requiring the generation of superoxides (produced by reduction of oxygen) [39].

Shestakova et al. [40] synthesised Ti–Ta$_2$O$_5$, Ti–SnO$_2$ and Ti–Ta$_2$O$_5$/SnO$_2$ electrodes and tested their effectiveness in photoelectrocatalytic degradation of methylene blue dye. The Ti–Ta$_2$O$_5$/SnO$_2$ demonstrated much higher photoelectrocatalytic activity than Ti–Ta$_2$O$_5$ and Ti/SnO$_2$ [40]. The outstanding photoelectrochemical activity of the Ti–Ta$_2$O$_5$/SnO$_2$ electrodes was ascribed to the enhanced mobility and separation of generated electrons and holes promoted by the photoinduced potential difference created at the Ti–Ta$_2$O$_5$/SnO$_2$ junction interface. Degradation of phenol in aqueous medium, and the treatment of paper mill industrial effluent has been performed using TiO$_2$/RuO$_2$ based electrode [41]. The result showed that the synergistic effects of electrolysis and photocatalysis of the photoelectrochemical process caused the improved performance of the TiO$_2$/RuO$_2$ heterostructure compared to the performance of the individual electrolysis and photocatalysis processes. Again, photoelectrocatalytic degradation of tetrabromobisphenol A was performed through the use of a graphene oxide modified cerium and titanium oxides nanotube arrays (CeO$_2$–TiO$_2$/rGO) based photoanode with a successful result [42]. The holes generated during light illumination were identified to play significant roles in the degradation process. CuS–GeO$_2$–TiO$_2$ and BiVO$_4$/TiO$_2$/Ti composites electrodes were also prepared by Wen and Zhang [43] and Hongxing et al. [44], respectively. The individual electrodes displayed impressive visible light and photoelectrocatalytic performance highlighting the usefulness of the mixed metal oxide system.

Fig. 9.4 Mechanism of trapping and prevention of electron-hole recombination by dopants

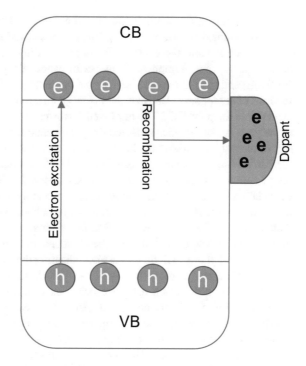

9.2.3 Doping with Metals and Non-metals

Doping is the introduction of impurities or foreign materials into metal oxide semi-conductor photocatalysts. Doping is normally intended to enhance the electronic, optical and overall efficiency of semiconductors. Doping causes improvement in the photocatalyst's efficiency through narrowing of its band gap and the resultant advancement in its visible light sensitivity, formation of impurity states and oxygen vacancies and improvement in its surface area. Doping also promotes rapping and prevention of electron–hole recombination [45, 46], where the empty orbitals of the dopants accept the excited electrons and prevent them from recombining with the holes (Fig. 9.4).

9.2.3.1 Doping with Metals

Doping of metal oxides with metals can be achieved by either substitutional or interstitial means. In substitutional doping, a host metal in the crystal lattice of the metal oxide is substituted by the dopant ion. Metal doping is considered to be interstitial when the dopant ion's radius is so small compared to that of the host metal that the dopant ion can enter the crystal lattice of the metal oxide [46].

The electrocatalytic degradation of aniline using a Ce-doped Ti/SnO$_2$–Sb electrode was studied by Xu et al. [47]. The result showed enhancement in the metal oxide's electrochemical properties after doping leading to about 98% aniline removal efficiency. In order to improve on its performance, TiO$_2$ was doped with platinum (Pt–TiO$_2$) and formed into an electrode for the removal of formic acid in water [48]. In terms of comparative analysis, the Pt–TiO$_2$ removed the pollutant as a higher efficiency than the pure TiO$_2$. The improved performance of the Pt–TiO$_2$ was attributed to the presence of Pt which attracted the photogenerated electrons and reduced their rate of recombination with the holes. For the degradation of a pharmaceutical waste (carbamazepine) in water, Ghasemian et al. [49] fabricated an electrode consisting of antimony doped tin (II) oxide-tungsten trioxide (Sb-doped SnO–WO$_3$). The degradation was done through photolytic, photocatalytic and photoelectrochemical processes and the results were compared with emphasis on the formation of intermediate products. The result revealed that fewer intermediate products were detected in the water sample treated through the photoelectrochemical process than those of photolytic and photocatalytic processes, indicating the superiority of the photoelectrochemical process in removing carbamazepine from polluted water over the other methods. The ability of an electrode based on lead doped-lead oxide (Pb–PbO$_2$) nanomaterial to effectively degrade disperse dye, reactive dye, and direct dye has also been studied by Morsi et al. [50]. The study was performed on the basis of the influence of certain factors including pH, reaction time, temperature and the initial concentrations of the dyes. The process was noted to be most effective at pH 3, 30 °C and 30 mA/cm^2 in the presence of NaCl (3 g/l) due to indirect oxidation of the dyes by hypochlorite ions produced through oxidation of chloride as well as direct oxidation by electrogenerated OH· adsorbed on the surface of the PbO$_2$.

9.2.3.2 Doping with Non-metals

It has been noted that non-metal doping results in much more efficient metal oxide semiconductors than metal doping as far as their stability and photocatalytic activity are concerned. Non-metal doping has also been identified to be less complex compared to metal doping [51]. Some of the non-metals mainly used as dopants of metal oxides include carbon, fluorine, sulphur, nitrogen, boron and phosphorus [52]. Upon doping, the non-metals introduce or create impurity energy levels near the conduction band of the metal oxide semiconductor photocatalyst [53]. The degree to which non-metal doping improves the efficiency of the catalyst relies on the electronic properties and relative size of the dopant [54].

A photoanode made up of boron-doped diamond known to possess high electrochemical activity has been prepared and deployed in the treatment of synthetic ternary wastewater, and removal of benzoic acid from water with excellent results [55]. Castellanos-Leal et al. [56] performed nitrogen and fluorine co-doping of TiO$_2$ and applied the catalyst in a three-electrode cell for the degradation of cyanide from water. They noticed that the rate at which the pollutant was degraded depended on the synergistic influence of the dopants and the pollutants concentrations. While doping with nitrogen was detected to create localised state leading to TiO_2 band gap reduction and the ensuing heightened visible light activity, the presence of fluoride was noted to promote the superficial properties and improved charge transfer mechanism of TiO_2 [56]. Liu et al. [57] also identified that upon doping TiO_2 with fluorine, the visible light sensitivity and photoelectrochemical performance of TiO_2 based photoanode increased significantly when it was applied to remove methylene blue dye from water. The performance of the electrode depended on fluorine dopant concentration and the pH of the medium. A high value of 97.8% dye removal rate was attained after 4 h at pH 9.94 and 1.4 V bias. Efficient separation of the charge carriers and the generation of sufficient hydroxyl radical where assigned to the impressive performance of the electrode. In addition, carbon doped TiO_2 based electrode fabricated by Calva-Yáñez et al. [58] demonstrated impressive visible light activity. About 20 times photocurrent enhancement and lower electron magnitude of about two orders were displayed by the fabricated carbon-doped TiO_2 based electrode compared to the undoped TiO_2 electrode. The improvement in photocurrent after carbon doping indicates the valuable role played by carbon by increasing the surface area of TiO_2 and improving the charge transport tendencies which influences the efficiency of organic pollutants degradation through photoelectrochemical degradation process [58]. Nitrogen-doped TiO_2 based electrode produced by Sakthivel et al. [59] also demonstrated a notable charge carrier transport, improved surface area, and photoelectrocatalytic properties.

9.2.3.3 Metal and Non-metal Co-doping

Metal and non-metal co-doping of metal oxide semiconductor photocatalysts has been adopted as a suitable technique for harnessing the synergistic properties of metal-doped, and non-metal doped semiconductors in order to achieve augmented photoelectrochemical properties. In this technique, the metal and non-metal dopants are concurrently introduced into the host metal oxide [60]. Co-doping improves the general performance of the catalysts by changing their photoactivity from ultraviolet to visible light absorption through narrowing of the catalyst's band gap. It also results in enlargement of the catalyst's surface area through particle size reduction for enhanced adsorption of the pollutants. In addition, co-doping reduces the rate of recombination of the photogenerated electrons and hole [61] leading to augmentation of pollutants removal process through the use of photoanodes under light illumination.

A number of research works have been carried out on catalyst co-doping with metals and non-metals. The results of these works evidenced the profound photoelectrochemical properties of these kinds of modified metal oxides in comparison to individual metal doped and non-metal doped semiconductors. In their mission to evaluate the electrocatalytic tendencies of metal and non-metal co-doped metal oxides, Yan et al. [62] fabricated cobalt and boron co-doped TiO_2 nanotube array (Co, B–TiO_2) and deployed it in the degradation of formaldehyde and oxytetracycline under visible light irradiation. According to their result, the Co, B–TiO_2 based electrode displayed a better pollutants removal tendency than the pure TiO_2 and the individual $Co–TiO_2$ and $B–TiO_2$ based electrodes. A copper and nitrogen co-doped TiO_2 based electrode meant to remove humic acid has also been prepared [63]. The undoped TiO_2 degraded about 41% of humic acid in 219 min while a degradation efficiency of about 93% was achieved when the electrode was applied to degrade humic acid upon simultaneous application of visible light and 5 V anodic bias potential within the same time. While nitrogen doping was noted to improve visible light responsivity and promoted quantum yield, the copper dopant was noted to trap the electrons and reduced their recombination rates with the holes resulting in the excellent performance of the electrode [63].

Similarly, cerium (Ce) and boron (B) co-doped TiO_2 nanocomposite (Ce B–TiO_2) based electrode prepared by Wang et al. [64] showed a high photoelectrocatalytic potential under visible light. The high photoelectrocatalytic performance of the Ce B–TiO_2 based electrode was credited to lowing of the rate of recombination of the electrons and holes, and band gap lowering. The ability of a copper and nitrogen co-doped titanium dioxide photoanode to treat landfill leachate has been studied with remarkable results [65]. Bias potential, pH and initial chemical oxygen demand were some of the parameters taken into consideration during this treatment process. A pH of 10, COD of 4377.98 mg L^{-1} and potential bias of 25 V were the conditions at which the optimum treatment was achieved [65].

9.3 Conclusion

The environmental and health effects posed by the use of unclean or polluted water and the significance of treating wastewater before discharging it into the environment have been highlighted in this chapter. Discussion on technological developments aimed at resolving water pollution problems has also been made with an emphasis on the identification of a suitable technique for the proficient degradation of organic water pollutants. Photoelectrocatalysis was recognized as an environmentally friendly and cost-effective method for the elimination of organic pollutants from water, and the limitations associated with the use of metals oxides in electrophotocatalytic removal of organic pollutants from water was discussed. Some of the modifications that have been made purposely to enhance the efficiency of metal oxide semiconductors in organic pollutants degradation from water through photoelectrocatalysis, including modification with reduced graphene oxide, exfoliated graphite, carbon nanotube, doping with metals and non-metals, and co-doping with metals and non-metals were reviewed.

References

1. T. Robinson, G. McMullan, R. Marchant, P. Nigam, Remediation of dyes in textile effluent: a critical review on current treatment technologies with a proposed alternative. Biores. Technol. **77**(3), 247–255 (2001)
2. P.H. Gleick, A look at twenty-first-century water resources development. Water Int. **25**(1), 127–138 (2000)
3. S.H. Chan, WuT Yeong, J.C. Juan, C.Y. Teh, Recent developments of metal oxide semiconductors as photocatalysts in advanced oxidation processes (AOPs) for treatment of dye waste-water. J. Chem. Technol. Biotechnol. **86**(9), 1130–58 (2011)
4. H.Y. Shu, M.C. Chang, H.H. Yu, W.H. Chen, Reduction of an azo dye Acid Black 24 solution using synthesized nanoscale zerovalent iron particles. J. Colloid Interface Sci. **314**(1), 89–97 (2007)
5. M.I. Litter, Introduction to photochemical advanced oxidation processes for water treatment, in *Environmental Photochemistry Part II* (Springer, Berlin, 2005), pp. 325–366
6. B. Ntsendwana, S. Sampath, B.B. Mamba, O.A. Arotiba, Photoelectrochemical oxidation of p-nitrophenol on an expanded graphite–TiO_2 electrode. Photochem. Photobiol. Sci. **12**(6), 1091–1102 (2013)
7. F. Chen, F. Yan, Q. Chen, Y. Wang, L. Han, Z. Chen, S. Fang, Fabrication of Fe_3O_4–SiO_2–TiO_2 nanoparticles supported by graphene oxide sheets for the repeated adsorption and photocatalytic degradation of rhodamine B under UV irradiation. Dalton Trans. **43**(36), 13537–13544 (2014)
8. J. Li, S. Lv, Y. Liu, J. Bai, B. Zhou, X. Hu, Photoeletrocatalytic activity of an n-ZnO/p-Cu_2O/n-TNA ternary heterojunction electrode for tetracycline degradation. J. Hazard. Mater. **262**, 482–488 (2013)
9. H.J. Choi, S.M. Jung, J.M. Seo, D.W. Chang, L. Dai, J.B. Baek, Graphene for energy conversion and storage in fuel cells and supercapacitors. Nano Energy **1**(4), 534–551 (2012)
10. K.S. Siddhapara, D.V. Shah, Experimental study of transition metal ion doping on TiO_2 with photocatalytic behavior. J. Nanosci. Nanotechnol. **14**(8), 6337–6341 (2014)
11. Y. Zhu, F. Piscitelli, G.G. Buonocore, M. Lavorgna, E. Amendola, L. Ambrosio, Effect of surface fluorination of TiO_2 particles on photocatalytic activity of a hybrid multilayer coating obtained by sol-gel method. ACS Appl. Mater. Interfaces. **4**(1), 150–157 (2011)
12. S.J. Moniz, S.A. Shevlin, D.J. Martin, Z.X. Guo, J. Tang, Visible-light driven heterojunction photocatalysts for water splitting—a critical review. Energy Environ. Sci. **8**(3), 731–759 (2015)
13. L. Pan, J.J. Zou, X. Zhang, L. Wang, Water-mediated promotion of dye sensitization of TiO_2 under visible light. J. Am. Chem. Soc. **133**(26), 10000–10002 (2011)
14. J.N. Schrauben, R. Hayoun, C.N. Valdez, M. Braten, L. Fridley, J.M. Mayer, Titanium and zinc oxide nanoparticles are proton-coupled electron transfer agents. Science **336**(6086), 1298–1301 (2012)
15. Y. Bai, H. Yu, Z. Li, R. Amal, G.Q. Lu, L. Wang, In situ growth of a ZnO nanowire network within a TiO_2 nanoparticle film for enhanced dye-sensitized solar cell performance. Adv. Mater. **24**(43), 5850–5856 (2012)
16. A.S. Mayorov, R.V. Gorbachev, S.V. Morozov, L. Britnell, R. Jalil, L.A. Ponomarenko, A.K. Geim, Micrometer-scale ballistic transport in encapsulated graphene at room temperature. Nano Lett. **11**(6), 2396–2399 (2011)
17. K.S. Novoselov, A.K. Geim, S. Morozov, D. Jiang, M. Katsnelson, I. Grigorieva, A.A. Firsov, Two-dimensional gas of massless Dirac fermions in graphene. Nature **438**(7065), 197 (2005)
18. Y. Guo, K. Xu, C. Wu, J. Zhao, Y. Xie, Surface chemical-modification for engineering the intrinsic physical properties of inorganic two-dimensional nanomaterials. Chem. Soc. Rev. **44**(3), 637–646 (2015)
19. X.H. Zhou, L.H. Liu, X. Bai, H.C. Shi, A reduced graphene oxide based biosensor for high-sensitive detection of phenols in water samples. Sens. Actuators B: Chem. **181**, 661–667 (2013)
20. E.H. Umukoro, M.G. Peleyeju, J.C. Ngila, O.A. Arotiba, Photoelectrochemical degradation of orange II dye in wastewater at a silver–zinc oxide/reduced graphene oxide nanocomposite photoanode. RSC Adv. **6**(58), 52868–52877 (2016)

21. F.C. Moraes, L.F. Gorup, R.S. Rocha, M.R. Lanza, E.C. Pereira, Photoelectrochemical removal of 17β-estradiol using a RuO_2-graphene electrode. Chemosphere **162**, 99–104 (2016)
22. M. Wang, X. Shang, X. Yu, R. Liu, Y. Xie, H. Zhao et al., Graphene–CdS quantum dots-polyoxometalate composite films for efficient photoelectrochemical water splitting and pollutant degradation. Phys. Chem. Chem. Phys. **16**(47), 26016–26023 (2014)
23. Q. Shi, X. Song, H. Wang, Z. Bian, Enriched photoelectrochemical performance of phosphate doped $BiVO_4$ photoelectrode by coupling FeOOH and rGO. J. Electrochem. Soc. **165**(4), H3018–H3027 (2018)
24. P. Ramesh, G.S. Suresh, S. Sampath, Selective determination of dopamine using unmodified, exfoliated graphite electrodes. J. Electroanal. Chem. **561**, 173–180 (2004)
25. B. Ntsendwana, B.B. Mamba, S. Sampath, O.A. Arotiba, Synthesis, characterisation and application of an exfoliated graphite–diamond composite electrode in the electrochemical degradation of trichloroethylene. RSC Adv. **3**(46), 24473–24483 (2013)
26. O.M. Ama, Exfoliated graphite/tungsten trioxide nanocomposite electrode for the photoelectrochemical degradation of eosin yellow and methylene blue in wastewater. Int. J. Sci. Res. Methodol. Hum. **8**(3), 23–38 (2017)
27. O.M. Ama, K. Khoele, W.W. Anku, S.R. Suprakas, Photoelectrochemical degradation of 4-nitrophenol using CuOZnO/exfoliated graphite nanocomposite electrode. Int. J. Electrochem. Sci. **14**(3), 2893–2905 (2019)
28. X. Yu, L. Qiang, Preparation for graphite materials and study on electrochemical degradation of phenol by graphite cathodes. Adv. Mater. Phys. Chem. **2**(02), 63 (2012)
29. N.G. Sahoo, S. Rana, J.W. Cho, L. Li, S.H. Chan, Polymer nanocomposites based on functionalized carbon nanotubes. Prog. Polym. Sci. **35**(7), 837–867 (2010)
30. Y. Cong, X. Li, Y. Qin, Z. Dong, G. Yuan, Z. Cui, X. Lai, Carbon-doped TiO_2 coating on multiwalled carbon nanotubes with higher visible light photocatalytic activity. Appl. Catal. B **107**(1–2), 128–134 (2011)
31. B.K. Vijayan, N.M. Dimitrijevic, D. Finkelstein-Shapiro, J. Wu, K.A. Gray, Coupling titania nanotubes and carbon nanotubes to create photocatalytic nanocomposites. Acs Catal. **2**(2), 223–229 (2012)
32. C.Y. Kuo, Prevenient dye-degradation mechanisms using UV/TiO_2/carbon nanotubes process. J. Hazard. Mater. **163**(1), 239–244 (2009)
33. F.W.P. Ribeiro, L.H. Mascaro, S.A. Alves, Visible light-induced photoelectrocatalytic degradation of 4-nitrophenol on $BiVO_4$/carbon nanotube electrode. In *Meeting Abstracts* (The Electrochemical Society, 2015), Vol. 30, pp. 1732–1732
34. D. Chaudhary, S. Singh, V.D. Vankar, N. Khare, ZnO nanoparticles decorated multi-walled carbon nanotubes for enhanced photocatalytic and photoelectrochemical water splitting. J. Photochem. Photobiol., A **351**, 154–161 (2018)
35. Ivana, C. (2019). Enhancement of photoelectrocatalysis efficiency of carbon nanotubes doped with TiO_2 nanostructures applied on pesticide degradation. Compos. Mater. Res. **6**(3)
36. Á. Jakab, A. Pop, C. Orha, F. Manea, R. Pode, Electrochemical degradation and determination of pentachlorophenol from water using TiO_2-modified zeolite-carbon composite electrodes. Environ. Eng. Manag. J. **13**(9), 2159–2165 (2014)
37. F.J. Zhang, M.L. Chen, W.C. Oh, Characterization of CNT/TiO_2 electrode prepared through impregnation with TNB and their photoelectrocatalytic properties. Environ. Eng. Res. **14**(1), 32–40 (2009)
38. Z. He, Y. Li, Q. Zhang, H. Wang, Capillary microchannel-based microreactors with highly durable ZnO/TiO_2 nanorod arrays for rapid, high efficiency and continuous-flow photocatalysis. Appl. Catal. B **93**(3–4), 376–382 (2010)
39. I. Bedja, P.V. Kamat, Capped semiconductor colloids. Synthesis and photoelectrochemical behavior of TiO_2 capped SnO_2 nanocrystallites. J. Phys. Chem. **99**(22), 9182–9188 (1995)
40. M. Shestakova, P. Bonete, R. Gómez, M. Sillanpää, W.Z. Tang, Novel Ti/Ta_2O_5–SnO_2 electrodes for water electrolysis and electrocatalytic oxidation of organics. Electrochim. Acta **120**, 302–307 (2014)

41. R.T. Pelegrini, R.S. Freire, N. Duran, R. Bertazzoli, Photoassisted electrochemical degradation of organic pollutants on a DSA type oxide electrode: process test for a phenol synthetic solution and its application for the E1 bleach kraft mill effluent. Environ. Sci. Technol. **35**(13), 2849–2853 (2001)
42. Q. Zhou, A. Xing, D. Zhao, K. Zhao, Tetrabromobisphenol A photoelectrocatalytic degradation using reduced graphene oxide and cerium dioxide comodified TiO_2 nanotube arrays as electrode under visible light. Chemosphere **165**, 268–276 (2016)
43. X. Wen, H. Zhang, Photoelectrochemical properties of $CuS–GeO_2–TiO_2$ composite coating electrode. PLoS ONE **11**(4), e0152862 (2016)
44. D. Hongxing, L. Qiuping, H. Yuehui, Preparation of nanoporous $BiVO_4/TiO_2/Ti$ film through electrodeposition for photoelectrochemical water splitting. R. Soc. Open Sci. **5**(9), 180728 (2018)
45. P. Sathishkumar, R.V. Mangalaraja, S. Anandan, M. Ashokkumar, $CoFe_2O_4/TiO_2$ nanocatalysts for the photocatalytic degradation of Reactive Red 120 in aqueous solutions in the presence and absence of electron acceptors. Chem. Eng. J. **220**, 302–310 (2013)
46. Y. Cao, Y. Yu, P. Zhang, L. Zhang, T. He, Y. Cao, An enhanced visible-light photocatalytic activity of TiO_2 by nitrogen and nickel–chlorine modification. Sep. Purif. Technol. **104**, 256–262 (2013)
47. H. Xu, A. Li, X. Cheng, Electrochemical performance of doped SnO_2 coating on Ti base as electrooxidation anode. Int. J. Electrochem. Sci. **6**, 5114–5124 (2011)
48. C. He, M. Abou Asi, Y. Xiong, D. Shu, X. Li, Photoelectrocatalytic degradation of organic pollutants in aqueous solution using a $Pt–TiO_2$ Film. Int. J. Photoenergy (2009)
49. S. Ghasemian, D. Nasuhoglu, S. Omanovic, V. Yargeau, Photoelectrocatalytic degradation of pharmaceutical carbamazepine using Sb-doped Sn80%–W20%-oxide electrodes. Sep. Purif. Technol. **188**, 52–59 (2017)
50. M.S. Morsi, A.A. Al-Sarawy, W.S. El-Dein, Electrochemical degradation of some organic dyes by electrochemical oxidation on a Pb/PbO_2 electrode. Desalin. Water Treat. **26**(1–3), 301–308 (2011)
51. M. Ghaffari, H. Huang, P.Y. Tan, O.K. Tan, Synthesis and visible light photocatalytic properties of $SrTi (1–x) FexO (3 − \delta)$ powder for indoor decontamination. Powder Technol. **225**, 221–226 (2012)
52. H. Wang, J.P. Lewis, Effects of dopant states on photoactivity in carbon-doped TiO_2. J. Phys.: Condens. Matter **17**(21), L209 (2005)
53. J.C. Yu, J. Yu, W. Ho, Z. Jiang, L. Zhang, Effects of F-doping on the photocatalytic activity and microstructures of nanocrystalline TiO_2 powders. Chem. Mater. **14**(9), 3808–3816 (2002)
54. M. Wang, J. Ioccozia, L. Sun, C. Lin, Z. Lin, Inorganic-modified semiconductor TiO_2 nanotube arrays for photocatalysis. Energy Environ. Sci. **7**(7), 2182–2202 (2014)
55. F. Montilla, G. Quijano, D. Alonso, E. Morallon, *Synthetic Boron-Doped Diamond Electrodes for Electrochemical Water Treatment* (2014)
56. E.L. Castellanos-Leal, P. Acevedo-Peña, V.R. Güiza-Argüello, E.M. Córdoba-Tuta, N and F codoped TiO_2 thin films on stainless steel for photoelectrocatalytic removal of cyanide ions in aqueous solutions. Mater. Res. **20**(2), 487–495 (2017)
57. D. Liu, R. Tian, J. Wang, E. Nie, X. Piao, X. Li, Z. Sun, Photoelectrocatalytic degradation of methylene blue using F doped TiO_2 photoelectrode under visible light irradiation. Chemosphere **185**, 574–581 (2017)
58. J.C. Calva-Yáñez, M.S. de la Fuente, M. Ramírez-Vargas, M.E. Rincón, Photoelectrochemical performance and carrier lifetime of electrodes based on MWCNT-templated TiO_2 nanoribbons. Mater. Renew. Sustain. Energy **7**(3), 19 (2018)
59. S. Sakthivel, M. Janczarek, H. Kisch, Visible light activity and photoelectrochemical properties of nitrogen-doped TiO_2. J. Phys. Chem. B **108**(50), 19384–19387 (2004)
60. M. Xing, Y. Wu, J. Zhang, F. Chen, Effect of synergy on the visible light activity of B, N and Fe co-doped TiO_2 for the degradation of MO. Nanoscale **2**(7), 1233–1239 (2010)
61. Y.F. Li, D. Xu, J.I. Oh, W. Shen, X. Li, Y. Yu, Mechanistic study of codoped titania with nonmetal and metal ions: a case of C + Mo codoped TiO_2. Acs Catalysis **2**(3), 391–398 (2012)

62. G. Yan, M. Zhang, J. Hou, J. Yang, Photoelectrochemical and photocatalytic properties of N + S co-doped TiO_2 nanotube array films under visible light irradiation. Mater. Chem. Phys. **129**(1–2), 553–557 (2011)

63. X. Zhou, Y. Zheng, D. Liu, S. Zhou, Photoelectrocatalytic degradation of humic acids using codoped TiO_2 film electrodes under visible light. Int. J. Photoenergy (2014)

64. P. Wang, M. Cao, Y. Ao, C. Wang, J. Hou, J. Qian, Investigation on Ce-doped TiO_2-coated BDD composite electrode with high photoelectrocatalytic activity under visible light irradiation. Electrochem. Commun. **13**(12), 1423–1426 (2011)

65. X. Zhou, S. Zhou, X. Feng, Optimization of the photoelectrocatalytic oxidation of landfill leachate using copper and nitrate co-doped TiO_2 (Ti) by response surface methodology. PLoS ONE **12**(7), e0171234 (2017)

Chapter 10
Metal Oxide Nanocomposites for Adsorption and Photoelectrochemical Degradation of Pharmaceutical Pollutants in Aqueous Solution

L. Mdlalose, V. Chauke, N. Nomadolo, P. Msomi, K. Setshedi, L. Chimuka, and A. Chetty

Abstract The global deterioration of water quality which is associated with industrialisation, urbanisation, and a growing population is reaching critical levels and thus needs to be addressed urgently. Common pollutants that are discharged from industries and sewage plants include unknown toxic chemicals, heavy-metals and micro-organisms; these are well known and thoroughly studied. Of growing and great concern to both human and animal health is the new emerging class of pollutants known as endocrine disruptor chemicals (EDCs) or emerging organic compounds (EOCs); these are frequently associated with residues from pharmaceutical industries, i.e. they comprise of common drugs such as antibiotics, medication for chronic illnesses, pain killers. Regrettably, the traditional water purification systems cannot fully remove these pollutants, thus they are found in various water systems in minute concentrations. The danger is in the long run accumulative exposure to humans, animals and the environment. There are several methods that have been developed, reported and used for the removal of these pollutants. Several removal or remediation technologies have been studied and reported for the mineralisation of these emerging organic pollutants and of interest to this work is photocatalysis using light harvesting materials such TiO_2 (i.e. semiconductors) and electrochemistry. The drawbacks associated with semiconductors are low quantum yields that emanate from rapid recombination of photo-generated electrons and holes with very low lifetimes. To overcome these drawbacks and to enhance degradation, an electrical external field can be applied across the catalyst or semiconductor to induce special separation of photo-generated electron hole pair to allow a sink for the electrons in a process

L. Mdlalose (✉) · V. Chauke · N. Nomadolo · K. Setshedi · A. Chetty
Polymers and Composites, Materials Science and Manufacturing, Council for Scientific and Industrial Research, Pretoria, South Africa
e-mail: lmdlalose1@csir.co.za

P. Msomi
Department of Applied Chemistry, University of Johannesburg, Johannesburg, South Africa

L. Chimuka
Molecular Sciences Institute, School of Chemistry, University of the Witwatersrand, Johannesburg, South Africa

© Springer Nature Switzerland AG 2020
O. M. Ama and S. S. Ray (eds.), *Nanostructured Metal-Oxide Electrode Materials for Water Purification*, Engineering Materials,
https://doi.org/10.1007/978-3-030-43346-8_10

called photoelectrochemistry. This chapter highlights the reported mineralisation of organic pollutants photoelectrochemistry using semiconductors; it also highlights the efficiency of photoelectrocatalysis when compared with photocatalysis alone.

10.1 Introduction

Pharmaceutical residues in the aquatic environment are described as refractory effluents produced by mainly chemical-synthetic and fermentation processes. These processes generate huge volumes of wastewater which usually contains spent solvents, recalcitrant organics, pharmaceutical residues and flushed water from preparatory practices [1]. Thus, they are described as effluents consisting of complex ingredients composed of barely decomposed organic compounds with high concentrations of chemical oxygen demand (COD) and toxicity [2]. Along with development of medicinal industry, antibiotics are progressively utilized for medicinal applications and are added to animal feeds to inhibit the spread of diseases and promote growth. However, the inappropriate use of antibiotics endangers the environment, more especially water resources due to their persistence, biological activity, chronic toxicity effect on ecosystems and propagation of antibiotic resistance in microbes [3].

Pharmaceuticals are categorized as potential bioactive chemicals in the environment but what is more ominous, in waterbodies, they are reflected as emerging pollutants and still remain unregulated [4]. These compounds and their bioactive metabolites are constantly introduced in aquatic environment and they are detected in trace amounts and are becoming pseudo-persistent. Seemingly, most urban wastewater treatment plants are contaminated with medicinal compounds [5]. Their persistence in the environment is attributed to incomplete elimination in sewage plant treatments with 60–90% residual after biodegradation, deconjugation, sorption and photodegradation treatment steps [5]. The occurrence and fate of pharmaceutical residues in the environment have been reviewed, where they are ubiquitous in soil and aquatic samples [6, 7].

The above concerns reflect the crucial attention and need for the complete elimination of pharmaceuticals and their metabolites from aquatic systems to hinder their prospective toxicity and harmful health effects. Bearing in mind that these pollutants cannot be totally demolished in sewage treatment plants with conventional techniques, alternative techniques with efficient oxidation ability even at trace organic pollutants are mandatory.

Over the years, decontamination and disinfection of aqueous solutions by means of direct or cohesive electrochemical processes are measured as appealing alternatives due to the substantial improvement of the electrode materials and the combination with low-cost renewable energy sources. Numerous electrochemical technologies are available and been reviewed for wastewater remediation of pharmaceutical residues from both synthetic solutions and real pharmaceutical effluents. Electrochemical oxidation is identified as an environmental friendly technology with efficient degradation with minimal energy consumption and secondary pollution [2]. Furthermore, the implementation of photocatalytic oxidation has been of both academic and industrial attention owing to its potential capability in wastewater treatment field since it was discovered [2]. The metal oxide catalysts including TiO_2, SnO_2, and ZnO have been widely explored for photocatalytic activities [8–10]. TiO_2 nanoparticles have shown outstanding performance for its high oxidation capacity, non-toxicity and chemical stability. TiO_2 photocatalysts have been investigated for a quarter of a century [2].

There are some drawbacks reported for organic solution waste degradation by photocatalysis process. Hitchman et al. 2002 identified low quantum yield resulting from rapid recombination of photo-generated electrons and holes with lifetimes between picoseconds and nanoseconds [11]. Therefore, the holes have restricted time to react with the target substance and the photo-energy ultimately converts to heat energy without proficient chemical reaction [11]. In order to reduce electron/hole recombination, an electric field can be supplied across the catalyst film to stimulate spatial separation of photo-generated electron/hole pairs with an external circuit permitting a sink for the electrons. The process of electrochemical application to photocatalysis is called photoelectrochemical method. This method is regarded to have higher efficiency for mineralisation of organic pollutants in wastewater remediation. This chapter presents and examines approaches and opportunities of metal oxide/composites for application in photochemical catalysis method coupled with electrochemical technology for environmental remediation. Important parallels and differences between recent methods and proposed mechanisms of improvement will be highlighted. Environmental sustainability of the method and future challenges are also the primary key area of this segment.

10.2   Bases of Photoelectrocatalytic Activities

Electrochemical advanced oxidation processes (EAOPs) have been widely explored for the removal of organic pollutants from effluents while taking cautions of their health-risk factors in the environment. The EAOPs efficiency is established based on the in situ production of reactive oxygen species (ROS) such as hydroxyl radical ($\cdot$OH). Photoelectrocatalysis (PEC) has appeared as an impending powerful EAOP by merging photocatalytic and electrolytic method. It has been identified as an exceptional practise of increasing the photocatalytic efficiency for degrading organic pollutants [12]. The technique consists of presenting a biasing potential into the photocatalytic process [2]. In the set-up, a biasing potential is introduced across a photo-anode where a catalyst is supported. The arrangement permits more effective separation of photogenerated charges, in that way increasing the lifetime of electron-hole pairs [13].

The basic photoelectrocatalytic process on retarding the recombination of electron-hole pairs involves electron ejection from the valance band of a semiconductor (fully filled) to the vacant conductive band generating a positively charged vacancy or hole [14]. The band gap is linked to the light irradiation used. Exposing the semiconductor to an irradiation with greater energy than that of its band gap result

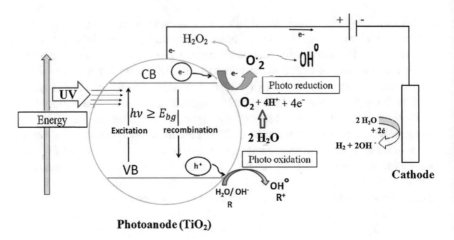

Photoanode (TiO₂)

Fig. 10.1 Mechanism for PEC process using TiO_2 photocatalyst. Adapted with permission from Daghrir et al. [14]

in photoexcitation of the electron in the conductive band from valance band to the conductive band. The light then permits the generation of electron-hole pairs where the hole in the valance band has strong oxidizing species and promoted electron in the conductive band is a possible redactor [14]. Organic pollutants are then oxidized by the photogenerated positively charged vacancy to complete mineralization.

A simplified photoelectrocatalytic process occurring at the surface of TiO_2 surface is shown in Fig. 10.1 [14]. The key feature that enhances photoelectrocatalytic efficiency is the potential that is applied [15]. The presentation in Fig. 10.1 shows that when a positive bias is applied to Ti/TiO_2 photoanode, the generated electrons can be moved into the external circuit instead of the oxygen molecule [14]. In the course, the photogenerated hole or hydroxyl radical will remain at the surface of the TiO_2 electrode. As a result, the rate of photogenerated electron-hole recombination is restricted whereas a probability exists to enhance the efficiency oxidation at the semiconductor-electrolyte interface. Other than the applied potential, there are several parameters that affect the conversion rate which include semiconductors specific feature, diffusion light, adsorption and desorption of reactants and products and the intensity of the electric field in the space charge region [16].

10.3 TiO₂ Metal Oxide Based Composites for Photoelectrocatalysis Enhancement

Semiconductor-based constituents with advanced nanostructures have become the fundamental element for photoconversion-correlated technology. TiO_2 is the most investigated photocatalyst even though its photocatalytic efficiency is still not up to the mark to meet the practical need under solar spectrum. Improved photo-catalytic processes have the potential in tackling some of society's utmost chal-lenges; most prominently in meeting energy demands and attacking environmen-tal pollution. Therefore, many efforts have been afforded to address TiO_2 bottle-necks which involve doping TiO_2 with metals/non-metals, decorating with narrow-band-gap semiconductors, combining with carbonaceous materials/polymers and so forth.

Despite the intrinsic band gap of TiO_2 limiting its optical absorption in the UV region, transforming TiO_2 powder to films results in significant decrease in surface area and hence reduced pollutant degradation efficiency [17]. The implementation of TiO_2 films is regarded practical compared to powdered catalyst to avoid high costs of TiO_2 powder separation post-wastewater treatment process [18]. To address the issue of poor degradation efficiency, photoelectrocatalyitic degradation which involve applying an external bias potential serve as an alternative for improving photocatalytic activities.

10.3.1   Transition Metals-Modified TiO₂

A number of research has been directed towards using transition metals and mostly tungsten as one of the best elements in narrowing the band gab of TiO_2 [17, 19]. These were made in the form of nanopowder [20], film [17] and core-shell [21]. Despite variations in the synthetic methods used in forming the mentioned products, these approaches have limitations. Liwera et al. 2011 fixed TiO_2 with tungsten oxide yet the catalyst was easily lost during photocatalysis [22]. When plasma electrolytic approach was used to fabricate tungsten laden TiO_2 nanotubes, a special appara-tus was mandatory to attain the spluttering coating [23]. Das et al. [24] achieved tungsten-doped TiO_2 nanotubes with alloys but came across difficulties in adjusting the dopant concentration [24]. A controllable approach was demonstrated by Gong et al. [25] when fabricating tungsten doped TiO_2 nanotubes through electrochemical oxidation of Ti substrate in glycerol/fluoride electrolyte having sodium tungstate [25]. The optimum effectiveness of the photoelectrocatalytic properties of tungsten doped TiO_2 was prepared by doping different sodium tungstate concentrations in the elec-trolyte [25]. The composite properties were also improved by annealing the electrode at various temperatures to achieve the anatase crystal phase. The great photoelectro-catalytic activities of the fabricated material were assessed through degradation of Rhodamine B and hydrogen generation [25].

Liquid phase deposition (LPD) technique was also successfully developed for preparing metal ion-doped TiO_2 films for visible light PEC degradation of dodecyl-benzenesulfonate (DBS) pollutant. The UV-vis spectra showed that doping the films with 1–7% content, the absorption wavelength was prolonged to visible light range and sensitive photocurrent response was observed under the illumination of visible light [17]). The optimal photocatalytic degradation of DBS of tungsten-TiO_2 film was 84.8% over 4 h treatment whereas applying +1.0 V anodic bias potential to visible light lead to 92% degradation over 90 min [17]. The degradation process was investigated in detail by systematically altering the functional bias potential, solution pH and the initial concentration affecting the degradation process.

On the bases of the combination of the TiO_2-tungsten composite material, it is crucial to note that the method of modifier deposition employed and the amount deposited can exert a much stronger effect on the PEC performance of the modified TiO_2 nanotubes [26]. In particular, a disproportionate quantity of tungsten or tungsten oxide can enhance charge recombination in the bicomponent, in that way reduce photocatalytic activity [27].

Martins and co-workers investigated a suitable amount of tungsten oxide that should be electro deposited of the surface of TiO_2 nanotubes for photoelectrochemical oxidation of the endocrine disruptor, propyl paraben (PPB) [28]. The photocurrent density of the tungsten decorated TiO_2 nanotubes showed excellent activity and durability which was 20% higher than that of pristine TiO_2 nanotube. Additionally, using the modified electrode resulted in complete removal of PPB within 20 min of photoelectrocatalysis performance at a solution pH of 3.0 and the applied voltage of 1.50 V. Complete mineralization was obtained in 60 min with 94% of total organic matter (TOC) removal [28]. The improved photoactivity maintained its stability even after continuous cycling between potentials.

In manipulating photocatalytic reactions by electrochemical methods, the external anodic bias potential on the irradiated metal-laden TiO_2 film does not only drive away the accumulated photogenerated electrons on metal particles to another cell, reducing the electron-hole recombination but can spatially isolate the capture of conduction band electrons from the oxidation process [29]. Investigations on hybrid technology associated with comprising noble metal deposition with the application of external field to improve the competency of photocatalytic degradation of organic pollutants have been explored. Li et al. [29] approach mainly devoted on photoelectrochemical performance silver-TiO_2 film supported on indium-tin oxide glass by a dip-coating and later photodeposition process [30]. The feasibility of silver deposited technology was also assessed in the presence of the external electric field. The findings demonstrated that the composite had an apparent additive effect with regard to suppressing the recombination amongst the photogenerated charge carriers while improving the photoelectrocatalytic oxidation of formic acid pollutant, which was chosen because it gets oxidized to carbon oxide without forming ant intermediates [30].

All the mentioned studies have indicated that loading a semiconductor with transition metals create new energy levels and treatment efficiencies in photoelectrocatalysis systems are more effective in degrading organic pollutants than photocatalysis alone. When comparing transition metals beneficial effects, platinum has been regarded as one of most effective metals in transferring photogenerated electron to reducible species at the catalyst surface [31]. Its effectiveness is not only measured based on only decreasing the occurrence of photogenerated electron-hole combination but the enhancement of oxygen reduction which has been identified as a potential rate-limiting step in the photocatalytic oxidation of organics [32]. However it has been demonstrated that TiO_2 platinization effectiveness in photoelectrocatalytic systems is also governed by the nature of organics being oxidized and the oxidation state of the photodeposited platinum [33]. Lee and Choi demonstrated that platinum zero is the most active specie of Pt deposits [34] whereas Teoh et al. postulated that there is a complex interrelationship between the Pt oxidation state and the degradation reaction of the organic matter [33].

To further elucidate the mechanistic pathway of TiO_2 platinization, the different processes between photocatalysis (PS) and photoelectrocatalysis (PES) were of great interest. In the former process, oxidation and reduction half reactions take place at different locations on the same TiO_2 crystal or particle [35]. The reduction half reaction will therefore be driven by the existence of electron acceptors to transfer the photogenerated electrons from the conductive band. This is often a controlling step of the overall reaction owing to low concentration of accessible electron acceptors near the surface [35]. As a result, surface deposited Pt could act as a catalyst to ease the reduction half reaction resulting enhanced overall efficiency. However in the PEC process, oxidation and reduction half reactions transpire separately at the anode and cathode, respectively [36]. As a result, the PEC process does not depend on the electron acceptor concentration (ref development of a direct). This is because the functional bias potential operates as an external driving force to transfer the photogenerated electrons from the conductive band to the external circuit then to the cathode where the forced half reactions occur [36]. Meaning the photo-consumption in a PEC method is different to that of a PC process and hence the impact of deposited Pt on a TiO_2 surface in PEC process is likely to be different to that of a PC process.

Gan et al. further showed that with low anodic potential bias of +0.1 V, the Pt deposited TiO_2 films performed way better than undoped-TiO_2 for oxidation of glucose [35]. Conversely, increasing the applied potential bias to > +0.5 V resulted more holes for oxidation of glucose on the TiO_2 surface in comparison to that of Pt–TiO_2 films. Under high applied anodic bias, Pt deposits may have detrimental result on the performance of a PEC system, for they could possibly block the TiO_2 surface (more especially with high Pt loading) and reduce the number of surface active sites [35].

While still on noble metal compounds, Ag/AgCl–TiO_2 composite has also been considered to be an efficient photocatalyst driven by the visible light [37]. Even though silver halides have been widely used for photographic films, they are uncertainly used as photocatalyst because of their instability in the visible light and high cost [38]. Implementation of surface plasmon resonance (SPR) effect of Ag/AgX makes it viable to fabricate an active and firm photocatalyst combining the advantages of Ag/AgCl nanoparticles with TiO_2 [39]. Liao et al. fabricated Ag/AgCl–TiO_2 nanaotube arrays electrode through a two-step method involving electrochemical anodization and electrodeposition processes for PEC degradation of microcystin-LR (MC-LR) [40]. MC are a result of cynobacteria found in lakes, rivers, ponds and portable water and are recognized as fresh water contaminating material [41]. They are extremely stable under visible light and because of their cyclic structures, they resist to decompose even at high temperatures and UV irradiation [42]. The mechanism of PEC degradation of the MC-LR was examined using different scavengers and it was discovered that the major functional species generated in situ were the holes, hydroxyl radicals and super-oxide radical anions [40]. Degradation of MC-LR

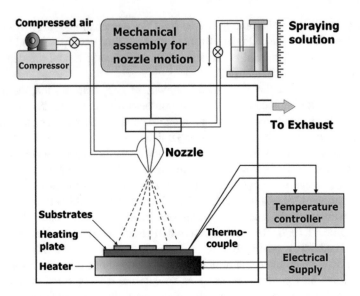

Fig. 10.2 Schematic set-up for spray pyrolysis method [45]

highly improves in sodium chloride electrolyte and in acid medium. The optimized Ag/AgCl–TiO$_2$ nanotube electrode displayed about 92% degradation and decreased to 75% in the presence of total organic carbons to 77.8% for a five hour reaction.

Among all metals, gold has been identified the top noble metal in producing the highest Schotty barrier potential [43]. Hence using gold as a dopant extends the photocatalytic response into the visible region as well as its rate [44]. The ability of gold doped TiO$_2$ films has been explored for photoelectrodegradation of benzoic acid [45]. Comparison studies of pristine TiO$_2$ and gold doped TiO$_2$ synthesized by chemical spray pyrolysis technique were investigated for PEC degradation [45]. The fundamentals of spray pyrolysis technique involve pyrolytic decomposition of a chosen compound to be deposited as shown in Fig. 10.2 [45]. In summary, the sprayed droplets reaching on the surface of the hot substrate experiences pyrolytic decomposition to form a single crystalline or bunch of crystallites as a final product while volatiles and solvents escape in the vapor phase. This process is regarded simple, cheap and entail uniform coverage of the film surface. Mohite et al. discovered the films were nanocrystalline in nature with tetragonal crystal structure using this technique. Further, for benzoic acid photocatalytic degradation, 49% improvement using Au/Ti atomic ratio of 3% than pristine TiO$_2$ films and Au doped films restored its degradation integrity for five cycles without post-treatment.

Membrane filtration technology has been investigated as a unique mineralization process under UV irradiation and voltage supply. The process was attained through deposition of graphite carbon and TiO$_2$ nanoparticles consecutively on aluminium oxide (Al$_2$O$_3$) membrane matrix to form a TiO$_2$/carbon/Al$_2$O$_3$ product [46]. The schematic diagram presented in Fig. 10.3 portray membrane photoelectrocatalytic

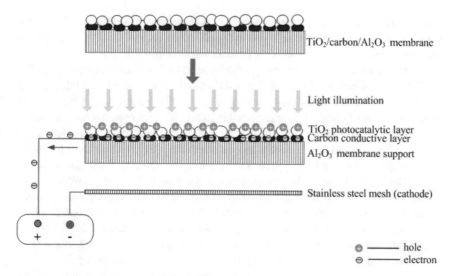

Fig. 10.3 Schematic diagram of photoelectrocatalytic membrane [46]

function when light is irradiated on the membrane then applying positive voltage on the membrane carbon layer and continuous transfer of electron to the cathode through the external circuit. Coupling membrane filtration with photoelectrocatalysis exhibited exceptional advantage on in Rhodamine B removal and antifouling capacity in the presence of natural organic matters [46]. The filtrate measured using Liquid chromatography–mass spectrometry LC/MS post treatment showed rapid reduction of Rhodamine B peak pollutant in the PEC process while three new intermediates chromatographic peaks were observed in the PC process.

10.3.2   Non-metals/Metalloids Modified TiO$_2$

Doping TiO$_2$ with non-metallic elements has been discovered as another alternative approach to facilitate light absorption of TiO$_2$ as well as restraining the recombination of photogenerated charge carriers. For example, doping TiO$_2$ with fluorine (F) results in the formation of reduced Ti^{3+} ions owing to the charge compensation between fluoride (F–) and Ti^{4+}, which restrain the recombination of photogenerated electron/hole pairs [47, 48]. Additionally, it enhances the surface acidity of TiO$_2$ encouraging the adsorption of organic pollutants [49].

Sol-gel method for preparation of F–TiO$_2$ particles has been widely used due to its simplicity and mildness in preparation and successful yield of the required product [47, 50, 51]. To assess the PEC performance, the prepared F–TiO$_2$ particles were attached onto the surface of F-doped tin oxide (SnO$_2$) glass by a screen-printing method to form F–TiO$_2$ photoelectrodes. Liu et al. evaluated the PC and PEC performance of the F–TiO$_2$ photoelectrodes using phenol as an analyte [48]. The findings demonstrated in Fig. 10.4 indicate performance evaluation of the two processes at different biases where EC degradation gradually increased until the optimum at applied 1.8 V (72.7%) whereas PEC outperformed it with the optimum (91.4%) degradation at 1.5 V. The poor reactivity with increase in the applied potential is attributed to the formation of more phenol intermediates which accumulates onto the photoelectrode surface hindering further access of phenol and thus lowering the degradation rate [48]. The PEC degradation mechanism was predominantly ascribed to the formation of active species such as hydroxyl radical by synergistic effect of F and tin co-doping, even though the tin content had no significant effect on active species [48].

Qu and Zhao revealed that boron-doped diamond (BDD) anodes are advantageous compared to other electrode materials due to their firm background current, wide range of working potential and stable surface microstructure [52]. BDD electrode's effectiveness was recommended based on the fact that it exhibits p-types semiconductor properties. Keeping in mind that TiO$_2$ photocatalyst exhibits n-type semiconductor properties; recombination of free electrons and holes can be reduced using p-n junction resulting a hybrid electrode designated as photoelectrocatalytic diodes [52]. A facile electrodeposition method was fabricated to make boron-doped TiO$_2$ to evaluate the photoelectrocatalytic activity and degradation of phenol contaminant [53]. Characterization analysis revealed improved crystallinity and shift of absorption toward the visible region for boron-doped TiO$_2$. The PEC measurements for phenol degradation under full spectrum illumination revealed higher catalytic activity for the PEC process with improved reaction rate compared to direct photolysis [53].

Sulphur-doped TiO$_2$/Ti photoelectrodes prepared by anodization were also investigated as a photoelectrocatalyzer. Different species of sulphur were implemented and the optimum doped sulphur headed to more lattice cell expansion which boosted the separation efficiency of photoinduced electrons and cavities [54]. Among the

Fig. 10.4 EC and PEC degradation of phenol by F–TiO$_2$ photoelectrode at different biases after 5 h of visible light irradiation [48]

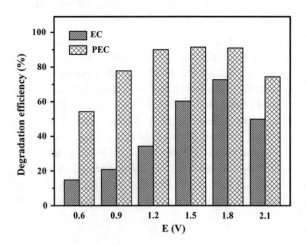

list of non-metals studied for doping, nitrogen seems to be more appealing due to its comparable atomic size with oxygen, metastable center formation and low ionization energy [55]. Moreover, the theoretical calculations projected an ideal band gap control for nitrogen doped TiO_2 and was in conformity with the experimental results [56]. Nitrogen-doped TiO_2 prepared via anodization and plasma based ion implantation showed response expansion of photoelectrodes towards visible light while diminishing the recombination of photo-generated holes and electrons [57]. Sun et al. recommended hydrothermal method to fabricate nitrogen-doped TiO_2 because it accelerates the interactions between solid and fluid species which eventually yields to a pure material [58]. The photoconversion efficiency and photoelectrocatalytic activity was certainly enhanced with the nitrogen-doped TiO_2 over pristine materials. In these cases PEC degradation performances were tested on organic dyes polluted solutions and not pharmaceutical waste.

10.3.3 Carbonaceous Materials Modified TiO_2

Carbon nanomaterials offer unique advantages due to their tunable structures, chemical inertness and electrical properties [59]. Additionally, carbon doping is favourable due to substantial overlap between doping states proximate the valance band edge and O2p state [59]. In particular, carbon nanotubes have extensively been used in heterogeneous photocatalysis due to large specific surface area, high adsorption capacities and being good dopants to support metal oxide nanoparticles [60]. Chemically altered carbon in conjunction with TiO_2 has been used to facilitate efficient photochemical splitting under illumination visible light and carbon doping results demonstrated higher photocatalytic activity with respect to unmodified TiO_2 [61]. To date, however, limited attention has been paid to PEC systems employing carbonaceous-TiO_2 composites as photoanodes under visible illumination for degradation of organic pollutants but there is a broader scope in photocatalysis.

Chen et al. synthesized carbon nanotubes-doped TiO_2 by modified sol-gel method and subsequently deposited the composite onto indium tin oxide conductive glass plate for phenol degradation. It was observed that carbon substituted oxygen in anatase lattice to form Ti–C and Ti–O–C bonds with Ti^{3+} species which was accountable for PEC activity under illumination with visible light [62]. It was prominent that carbon nanotubes content endorsed the transport of electrons to the substrate and increased the current however in excess it lowered TiO_2 crystallinity. Phenol effectively degraded through PEC over an appropriate carbon nanotubes content and with a TiO_2 electrode calcined at a moderate optimized temperature (400 °C) which led to elevated photocurrent [62].

Expanded graphite (EG) capabilities were also confirmed through modifying sol-gel strategy [50, 51]. TiO_2 anatase TiO_2 uniformly dispersed on the edge and interlayer of EG resulted Ti–O–C bonds. The fabricated EG-TiO_2 photoanode showed the 96.3% removal efficiency of phenol in 90 min at neutral solution pH in the PEC process. The proposed mechanism is demonstrated in Fig. 10.5 where there is high

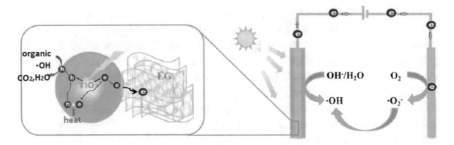

Fig. 10.5 Schematic representation of photoelectrocatalytic mechanism of EG/TiO$_2$ photoelectrode [51]

separation efficiency as generated electrons were transferred to the EG followed by applying external positive potential with electrons being migrated from anode to cathode through the inner circuit. The amassed electrons in the cathode and the outstanding photogenerated hole could simply react with the absorbed O$_2$, H$_2$O and OH—to yield hydroxyl radicals which are in control for the mineralization of organic pollutants [50, 51].

10.3.4 Polymer Modified TiO$_2$

Photoelectrode incorporated with polyaniline has been implemented in photoelectrochemical devices like solar cells [63]. Polyaniline has been a polymer of choice due to its good conductivity, oxidation-reduction capability, charge storage and photoelectric conversion performance [64]. Additionally, polyaniline is stable, easy to synthesize and effective charge transfer capabilities [64]. Therefore sensitizing TiO$_2$ with electronic performance of polyaniline is determined as an attractive approach to enhance the absorption and reflection of light in photocatalysis. Li et al. prepared polyaniline-TiO$_2$ nanotubes electrode via electrodeposition method for photoelectrocatalytic degradation of organic pollutants [65]. According to characterization techniques, the polyaniline-sensitized TiO$_2$ nanotube arrays presented a distinguishable red shift on the absorption spectrum and a strong interaction between the N–H groups on the polyaniline macromolecules and the TiO$_2$ nanotubes surface.

Yang and co-workers investigated the synergic effect of chromium and polyaniline components in modifying TiO$_2$ particles [66]. In this process, hexavalent chromium species was doped during the electrochemical oxidation synthesis procedure of TiO$_2$ nanotubes followed by adsorption of the polyaniline on the chromium doped nanotubes. The idea of incorporating different components was to improve the electron transition energy level of doped chromium transferring into TiO$_2$ conductive band while polyaniline absorbs light energy to encourage π–πn transition to move the excited state electrons in the valance band of TiO$_2$ to its conductive band [67, 68]. The photoelectrochemical performance of polyamine/chromium-TiO_2 nanotubes was investigated by voltammetry, photocurrent and electrochemical impedance spectroscopy. The photoelectrocatalytic degradation effectiveness of 10 mg/L p-nitrophenol was improved by 17.2% compared to chromium-TiO_2 nanotubes [66]. Under optimized conditions, 99.12% p-nitrophenol degraded with 75.46% removal of total organic carbons.

10.3.5   Ordered TiO₂ Materials

Photocatalytic efficiency of TiO_2 strongly depends on its unique nanostructure hence weak proficiency is achieved for bulk TiO_2 particles in a film. TiO_2 nanostructures such as tubes, wires dots, pillars and pillars have been investigated for light harvesting and photogenerated charge since electron transport is an essential factor for TiO_2 film performance [69, 70]. Among the TiO_2 nanostructured materials, tubular form has been considered most appropriate in improving the surface area without increasing the geometric area [71].

There are several method available for TiO_2 nanotubes preparation which include anodic oxidation, sol-gel, electrodeposition, hydrothermal and template synthesis [72, 73]. However the use of electrochemical anodization of Ti with fluorinated electrolytes is described as a relatively simple route for TiO_2 nanotube arrays fabrication. A typical anodization growth mechanism of TiO_2 nanotubes formation on a Ti substrate is shown in Fig. 10.6. Initially, a barrier layer on Ti surface is formed in the electrolyte from Ti oxidation step. As the layer breakdown under the influence of the applied electric field, nanopores are produced. By increasing the anodization time, anions in the electrolyte are able to migrate into the nanopores and Ti^{4+} species is generated from Ti oxidation and move from inside the nanopores to the electrolyte. Slower dissolution rate of the nanopores walls compared to compared to nanopores growth rate result in steady increase of diameter and length of the nanotubes. A build-up of anions follows during Ti–O compounds growth at the border triple of the nanopores ensuing in the production of soluble Ti–F and Ti–O–F compounds and subsequently the formation of nanotubes arrays [74].

The exceptional tubular architecture of TiO_2 nanotubes gives out high efficacy of electron transference and suppression of the recombination of photo-generated electron holes and the supply of external electric field assist the process even further. It has been investigated that on a saturated photocurrent density of TiO_2 nanotube photoanode, efficiency can be 23 times higher than TiO_2 nanofilm under similar conditions due to the existence of the cannula structure in nanotubes [74]. Ordered TiO_2 nanotube arrays with controllable inner-diameter, morphology and length were prepared via anodization method by adjusting electrolyte composition and anodization time [74]. PEC and PC activities were assessed against acyclovir pharmaceutical pollutant. A rapid degradation efficiency was obtained by PEC process compared to PC process with 97.1% for optimized photoanodes [74].

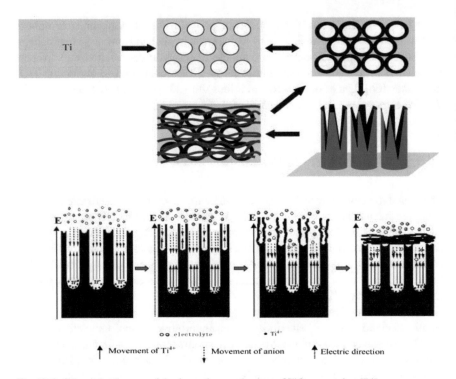

Fig. 10.6 Schematic diagram of the formation mechanism of TiO$_2$ nanotubes [74]

Zhang and co-workers confirmed that ordered TiO$_2$ nanotube arrays electrode being a more effective photoelectrode for achieving an improved organic pollutants degradation [71]. The exceptional performance was attributed to the highly ordered TiO$_2$ nanotube arrays structure which reduces the scattering of free electrons and heightens electron mobility. The anodization technique and sol-gel method were compared for ordered TiO$_2$ nanotube arrays and TiO$_2$ film electrode preparation and characterized using UV-vis absorption spectroscopy as indicated in Fig. 10.7.

Fig. 10.7 UV-Vis absorption spectra of **a** TiO$_2$ nanotube prepared with anodization technique and **b** TiO$_2$ film prepared with sol-gel method [71]

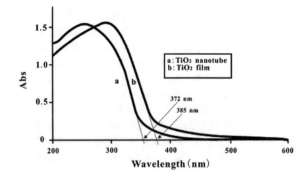

The UV-vis absorption band edge is known to be a strong function of the crystallite dimension of nanosize TiO_2 catalyst. A band gap absorption edges of 372 and 385 nm for TiO_2 nanotube and TiO_2 film, respectively. The 0.11 eV blue shift from TiO_2 film band gap energy (3.2 eV) signified insignificant TiO_2 particles formed [71].

Chlortetracycline (CTC) is one of the commonly used antibiotic or metabolite and has been detected in surface water, sewage water as well as ground water [75]. PEC process by means of UV light and ordered nanocrystalline TiO_2 photo-anodes were implemented for degradation of these pollutant [75]. The investigation included Ti/TiO_2 rectangular electrode preparation using the Pulsed Laser Deposition procedure. A rutile structure was obtained after coating TiO_2 at 600 °C with uniform thickness and uniformity. Most effective experimental parameters for CTC degradation were current density and treatment time. The effects of pollutant initial concentration and UV lamp position were only 4% and 2.5%, respectively. The study was further confirmed by means of central composite design and it was verified that current intensity and treatment time have positive responses while negative response was obtained on the removal of CTC [75]. The possible by-products and the mechanism of CTC degradation were not identified.

It has been mentioned numerously that PEC is a facile route to unravel fast recombination between photogenerated electrons and holes to improve the photocatalytic degradation efficiency of the highly ordered TiO_2 nanotube arrays and on how applying low bias potential significantly facilitate the transportation of photocarriers and suppress the recombination of photogenerated electrons and holes. According to Zhang and co-workers, the length of TiO_2 nanotubes and annealing temperature have strong influence on photocatalyisis whereas the external bias potential has insignificant effect [76]. Investigations showed the importance of prolonging the length of nanotubes to increase the surface area which is needed for photon absorption to ease the permeation of light. The optimum photoelectrocatalytic activity was below 8 μm of nanotubes length between 400 and 600 °C for annealing temperature. It was speculated that increasing the rutile amount somehow hindered the degradation of pollutants. Additionally, it was concluded that for photoelectrocalytic activities, the performance is not only influenced by the crystalline structure of the material but also closely associated with the micromorphology [76].

Liao et al. explained the importance of self-doped TiO_2 nanotube arrays to improve the photoelectrocatalytic activity of the material [77]. It was reported that metal doping TiO_2 with non-metals improve photoelectrocatalytic activity however, it hampers crystal and thermal stability and caused an in increase in carrier trapping which might result in poor performance of photocatalytic efficiency [78]. Self-doped TiO_2 nanotube arrays photoelectrodes were prepared and it was demonstrated that the photoelectrocatalytic responses were enhanced in both UV and visible regions, carrier density improved and reduction of charge transfer resistance between the interface and electrolyte were observed [77]. Seemingly, improving the photocatalytic activity of TiO_2 by self-doping techniques for UV and visible region is a challenging task for researchers [77]. The strategy was inspired by a hydrogenation method to generate disorder in the TiO_2 nanophase structure and improve visible light absorption of the material by Wang and co-workers [79].

10.4   Other Metal Oxide Catalysts Other Than TiO$_2$ Used for Photoelectrocatalytic Degradation

Photocatalytic materials using TiO$_2$ semiconductor is well established because of all its beneficial features. On the other hand, efforts have been paid to advance other photocatalytic materials that are active under sunlight. Metallic cations such as the 2 s configured trivalent bismuth ion (Bi^{3+}) in an oxide system is able to elevate the valance band by combining their respective orbitals with the O 2p orbital from oxygen in that way contracting the band gap of the semiconductor [80]. As a result oxides such as BiVO$_4$, Bi$_2$MoO$_6$ have been defined as promising photocatalysts for organic pollutant degradation [81, 80]. The photoelectrocatalytic properties of Bismuth molybdenum oxide (Bi$_2$MoO$_6$) deposited onto a boron-doped diamond surface were investigated using a dip-coating method and calcination [82]. The obtained hybrid electrodes exhibited effectual separation of the photoinduced charge carriers and clear enhancement in the photocurrent under visible light irradiation. It was intriguing that the fabricated hybrid electrode degraded both ibuprofen and naproxen via electro-oxidation and photocatalysis at an optimum applied bias potential of 2.0 V, however the chosen pollutants rapidly mineralize with photoelectrocatalytic process that the degradation rate was larger than the sum of photocatalysis and electro-oxidation processes [82]. The mineralization mechanism involved the opening of phenyl ring intermediates to form small molecular organic acids.

To improve photocatalytic activity of bismuth vanadate (BiVO$_4$) semiconductor, silicon was used to reduce the nanocrystal size leading to more available active sites and increase in surface charge carrier transfer rate [83]. The silicon-doped BiVO$_4$ film was synthesized by altered metalorganic decomposition method. The photoelectrocatalytic performance was assessed on phenol as a pollutant. The phenol degradation rate for silicon doped BiVO$_4$ electrode was 1.84 times greater that pristine BiVO$_4$ film electrode. The improvement was attributed to elevated surface hydrophilicity upon incorporating silica as well as decrease in crystal size as proven by contact angle and X-ray powder diffraction measurements [83].

Graphitic carbon nitride (g-C$_3$N$_4$) has been recently found as a suitable material to control BiVO$_4$ morphology and improve its photocatalysis potential. The uniqueness of g-C$_3$N$_4$ relies on its being a metal-free organic polymeric semiconductor with a band gap of 2.7 eV, hence it has attracted attention in photocatalytic water splitting [84]. Sun et al. fabricated g-C$_3$N$_4$-BiVO$_4$ composite on a fluorine-doped tin oxide (TFO) substrate as a photoanode for PEC degradation of diclofenac sodium (which is used as a non-steroidal anti-inflammatory drug). To increase diclofenan sodium degradation on the PEC system, hydrogen peroxide was added to activate photogenerated electrons at the cathode into the generation of hydroxyl radicals [85]. It was however interesting to note that degradation efficiency of diclofenan sodium pollutant using only hydrogen peroxide oxidant was negligible. The findings revealed that a concentration of 10 mM hydrogen peroxide was optimum to achieve the maximum degradation as superfluous concentrations could obstruct degradation by scavenging portions of reactive species [85].

The PEC oxidation process have been tested on polyaniline-intercalated layered manganese oxide nanocomposite electrode [86]. The electrode was facilely prepared by the delamination/reassembling process. As expected, the redox current of the composite film was much higher compared to the ones for individual components and hence PEC degradation was significantly accelerated on the polyaniline-manganese oxide film electrode [86].

Coupled semiconductors have been regarded as a great strategy to develop photo-electroactive anode materials and they offer improved degradation rates with greater efficacy for organic pollutants degradation. For example tin oxide (SnO_2) is electroactive while tungsten trioxide (WO_3) is regarded photoactive hence the combination of the two are selected as possible electrode materials used as anode for organic effluents disinfection [87, 88]. Ghasemian and Omanovic synthesized antimony-doped SnO_2 electrodes on a flat titanium substrate through a thermal deposition method for phenol degradation [89]. Pure SnO_2 had an optical band gap of 3.53 ± eV and its decreased to 2.56 ± 0.10 eV with 60% tungsten content but all the coated layers were photoactive under anodic bias representing the n-type semiconductivity nature. So the tin-doped WO_3 coated electrodes demonstrated efficient PEC activity for the degradation of phenol under UV light irradiation. It was however noted that the attained corresponding photocurrent was not fully dependant on the band-gap trend but there were other physico-chemical properties of metal-oxide coatings that influenced the degree of the photocurrent [89].

Tin-doped WO_3 anode material have also been explored for the removal of carbamazepine [90]. Carbazine is the utmost identified pharmaceutical in waste treatment plant effluents which is prescribed with approximately 1014 per year of global consumption [67, 68, 91]. The degradation rate and efficiency of carbamazepine increased with increasing applied current density until threshold value was reached. The expected acridine and acridone species known as transformation products with ecotoxicological properties were detected but were further oxidized in the PEC process at longer remedial times until complete mineralization was achieved with the 60 min treatment time. Once again, the UV-based remedial process in the absence of applied current density presented lower efficacy when matched with UV irradiation and electrical current in synergy. For carbamazepine removal, electrochemical oxidation played a major role that the photochemical oxidation process [90].

10.5   Adsorption Effects on Photoelectrocatalytic Performance of Photoelectrodes

It has been discovered that the photoelectrocatalytic oxidation process largely occurs on the electrode surface and not in the bulk solution, hence the relation concerning surface adsorption rate and photoreaction rate is imperative [92]. Generally, in the case where adsorption rate is faster than the photoreaction rate on the surface, the photoreaction rate would be a rate limiting factor for the observed photo-oxidation rate and vice versa. Further, adsorption is mainly driven by surface area, photo-electrode surface charge and the competing ions during the photoelectrocatalytic reaction. For example, sulphur doped TiO_2 photoelectrodes were facilely prepared as a way of endorsing the adsorption of surface hydroxyl during the photoelectrocatalytic performance [54]. The hydroxyl uptake was essential to capture photoinduced cavities and to form hydroxyl free radical which serves as a better oxidant to enhance oxidation reaction in the solution.

The effect of competing anions were tested against the photoelectrocatalytic activity using tungsten oxide decorated TiO_2 nanotubes photoelectrodes [93]. The chosen anions were chlorides, sulphates, bicarbonate, and nitrates where all adversely affected adsorption and degradation of 4-nonylphenol analyte in water. Chloride ions for example is known to be very mobile so it would rapidly compete and adsorb on the surface of tungsten oxide-TiO_2 nanotube photoelectrodes and get trapped on photogenerated holes leading to hydroxyl radicals turning hypochlorous acid which sabotages the degradation rate of 4-nonyphenol [94]. The presence of bicarbonate ions also tend to trap hydronium and give out carbon dioxide micro-bubble which adsorb on the photoelectrode surface which is likely to hinder light absorption. Additionally, carbon dioxide yield is also a photoelectrocatalysis process by product so it interrupts the chemical equilibrium of the degradation progression [93].

The generation of hydroxyl radical is as of equal importance to adsorption since it is frequently the primary oxidant for photoelectrocatalytic/photocatalytic reactions. This also correlates the solution pH which influences the photoelectrode surface charge density as well as the ionic analyte species present in the solution. When photoelectrocatalytic system was implemented on phenol degradation by carbon nanotubes-TiO_2 composite photoanodes, maximum phenol adsorption was observed at pH 3 however free radical generation was deprived resulting low photoelectrocatalysis competency [62]. Increasing the solution pH steered the increase in free hydroxyl radicals and positive PEC efficiency nevertheless at pH 11 minimal phenol could be adsorbed on the surface and hence the composite material was not sensitive enough for effective utilization of the hydroxyl free radicals therefore low PEC efficacy was obtained [62].

Lu and co-workers evaluated tetracycline hydrochloride- molecularly imprinted polymer (MIP) modified TiO_2 nanotube array electrode for enhanced photocatalytic activity and degradation of tetracycline hydrochloride (TC) [95]. The MIP modified electrode idea has high potential due to its surface area providing imprinted sites. Since its synthesis undergoes the combination of functional monomers and template, monomer polymerization with a crosslinker and removing the template; the MIP layer on the electrode showed high adsorption capability for the analyte (TC) [95].

10.6   Conclusions and Future Perspectives

From the turn of the century till today, various AOP processes which PEC is part of has drastically increased to overcome the occurrence of refractory species by improving its performance towards environmental remediation issues. In particular, treatment of effluents in association with electrochemical processes has been favoured for possibility of no waste production, chemicals free processes and greater ability to efficiently degrade pollutants. However, factors such as the complexity of the water matrix, the initial concentration of pollutants, the concentration of the oxidants and catalysts as well as the reactor configuration somehow limit its practicality and the maturity of the process. Therefore, following the great success obtained, energy and economic difficulties related to the process and its practicality are worth being investigated. The modularity, low specific footprint and easy automation make photoelectrocatalytic method to be most preferred amongst other advanced oxidation processes. However, it is worth mentioning that in some cases the by-products obtained after the treatment process can be more toxic compared to the initial analyte pollutant. These circumstances require more attention as they may require additional steps for further degradation. Similarly, the industrial scale application has not been fully evaluated as a result of economic difficulties and occasionally the lab-scale is still blurred, hence no scaling up is considered. In other cases the scaling-up process involve just the treatment of larger pollutant volume rather than the actual scale-up which include evaluating the impacts of the flow dynamics, materials as well as construction information.

Acknowledgements The authors acknowledge the Council for Scientific and Industrial Research (CSIR) and National Research Foundation (NRF), South Africa.

References

1. X. Shi, K.Y. Leong, H.Y. Ng, Bioresource technology anaerobic treatment of pharmaceutical wastewater: a critical review. Biores. Technol. **245**, 1238–1244 (2017)
2. T. Fang et al., Removal of COD and colour in real pharmaceutical wastewater by photoelectrocatalytic oxidation method. Environ. Technol. **34**, 779–786 (2013)
3. M. Ahmadi, H. Ramezani, N. Jaafarzadeh, Enhanced photocatalytic degradation of tetracycline and real pharmaceutical wastewater using MWCNT/TiO$_2$ nano-composite. J. Environ. Manage. **186**, 55–63 (2017)
4. I. Sirés, E. Brillas, Remediation of water pollution caused by pharmaceutical residues based on electrochemical separation and degradation technologies: a review. Environ. Int. **40**, 212–229 (2012)
5. O.A.H. Jones et al., Human pharmaceuticals in wastewater treatment processes. Crit. Rev. Environ. Sci. Technol. **35**, 401–427 (2005)
6. K. Ku, The presence of pharmaceuticals in the environment due to human use—present knowledge and future challenges. J. Environ. Manage. **90**, 2354–2366 (2009)
7. M.F.Ã. Rahman, E.K. Yanful, S.Y. Jasim, Occurrences of endocrine disrupting compounds and pharmaceuticals in the aquatic environment and their removal from drinking water: challenges in the context of the developing world. Desalination **248**, 578–585 (2009)
8. P.V. Laxma et al., TiO$_2$-based photocatalytic disinfection of microbes in aqueous media: a review. Environ. Res. **154**, 296–303 (2017)
9. A.M. Al-hamdi, U. Rinner, M. Sillanpää, Tin dioxide as a photocatalyst for water treatment. Process Saf. Environ. Prot. **107**, 190–205 (2017)

10. S. Bee et al., Photocatalytic water oxidation on ZnO: a review. Catalysts **7**, 93 (2017)
11. M.L. Hitchman, F. Tian, Studies of TiO_2 thin films prepared by chemical vapour deposition for photocatalytic and photoelectrocatalytic degradation of 4-chlorophenol. J. Electroanal. Chem. **538–539**, 165–172 (2002)
12. E. Zarei, R. Ojani, Fundamentals and some applications of photoelectrocatalysis and effective factors on its efficiency: a review. J. Solid State Electrochem. **21**, 305–336 (2017)
13. S. Garcia-segura, E. Brillas, Applied photoelectrocatalysis on the degradation of organic pollutants in wastewaters. J. Photochem. Photobiol., C **31**, 1–35 (2017)
14. R. Daghrir, P. Drogui, D. Robert, Photoelectrocatalytic technologies for environmental applications. J. Photochem. Photobiol., A **238**, 41–52 (2012)
15. N. Wang et al., Evaluation of bias potential enhanced photocatalytic degradation of 4-chlorophenol with TiO_2 nanotube fabricated by anodic oxidation method. Chem. Eng. J. **146**, 30–35 (2009)
16. H. Selcuk, J.J. Sene, M.A. Anderson, Photoelectrocatalytic humic acid degradation kinetics and effect of pH, applied potential and inorganic ions. J. Chem. Technol. Biotechnol. **78**, 979–984 (2003)
17. J. Gong et al., Liquid phase deposition of tungsten doped TiO_2 films for visible light photoelectrocatalytic degradation of dodecyl-benzenesulfonate. Chem. Eng. J. **167**, 190–197 (2011)
18. M. Pourmand, N. Taghavinia, TiO_2 nanostructured films on mica using liquid phase deposition. Mater. Chem. Phys. **107**, 449–455 (2008)
19. J. Gong et al., Tungsten and nitrogen co-doped TiO_2 electrode sensitized with Fe–chlorophyllin for visible light photoelectrocatalysis. Chem. Eng. J. **209**, 94–101 (2012)
20. P.G.W.A. Kompio et al., A new view on the relations between tungsten and vanadium in $V_2O_5WO_3/TiO_2$ catalysts for the selective reduction of NO with NH_3. J. Catal. **286**, 237–247 (2012)
21. W. Smith, Y. Zhao, Superior photocatalytic performance by vertically aligned core—shell TiO_2/WO_3 nanorod arrays. Catal. Commun. **10**, 1117–1121 (2009)
22. A. Lewera et al., Metal-support interactions between nanosized Pt and metal oxides (WO_3 and TiO_2) studied using X-ray photoelectron spectroscopy. J. Phys. Chem. C **115**, 20153–20159 (2011)
23. C.W. Lai et al., Preparation and photoelectrochemical characterization of WO_3-loaded TiO_2 nanotube arrays via radio frequency sputtering. Electrochim. Acta **77**, 128–136 (2012)
24. C. Das et al., Photoelectrochemical and photocatalytic activity of tungsten doped TiO_2 nanotube layers in the near visible region. Electrochim. Acta **56**, 10557–10561 (2011)
25. J. Gong et al., Novel one-step preparation of tungsten loaded TiO_2 nanotube arrays with enhanced photoelectrocatalytic activity for pollutant degradation and hydrogen production. Catal. Commun. **36**, 89–93 (2013)
26. A.A. Ismail et al., Ease synthesis of mesoporous WO_3–TiO_2 nanocomposites with enhanced photocatalytic performance for photodegradation of herbicide imazapyr under visible light and UV illumination. J. Hazard. Mater. **307**, 43–54 (2016)
27. L. Hinojosa-reyes, Solar photocatalytic activity of TiO_2 modified with WO_3 on the degradation of an organophosphorus pesticide. J. Hazard. Mater. **263**, 36–44 (2013)
28. A.S. Martin, et al., Enhanced photoelectrocatalytic performance of TiO_2 nanotube array modified with WO_3 applied to the degradation of the endocrine disruptor propyl paraben. J. Electroanal. Chem. **802**, 33–39 (2017)
29. Y. Li, G. Lu, S. Li, Photocatalytic hydrogen generation and decomposition of oxalic acid over platinized TiO_2. Appl. Catal. A **214**, 179–185 (2001)
30. C. He et al., Photoelectrochemical performance of Ag–TiO_2/ITO film and photoelectrocatalytic activity towards the oxidation of organic pollutants. J. Photochem. Photobiol., A **157**, 71–79 (2003)
31. H. Selcuk et al., Photocatalytic and photoelectrocatalytic performance of 1% Pt doped TiO_2 for the detoxification of water. J. Appl. Electrochem. **34**, 653–658 (2004)

32. C. Wang, A. Heller, H. Gerischer, Palladium catalysis of O_2 reduction by electrons accumulated on TiO_2 particles during photoassisted oxidation of organic compounds. J. Am. Chem. Soc. **114**, 5230–5234 (1992)

33. W. Yang, L. Mädler, R. Amal, Inter-relationship between Pt oxidation states on TiO_2 and the photocatalytic mineralisation of organic matters. J. Catal. **251**, 271–280 (2007)

34. J. Lee, W. Choi, Photocatalytic reactivity of surface platinized TiO_2: substrate specificity and the effect of Pt oxidation state. J. Phys. Chem. B **109**, 7399–7406 (2005)

35. W.Y. Gan, et al., A comparative study between photocatalytic and photoelectrocatalytic properties of Pt deposited TiO_2 thin films for glucose degradation. Chem. Eng. J. (2010), pp. 482–488

36. H. Zhao et al., Development of a direct photoelectrochemical method for determination of chemical oxygen demand. Anal. Chem. **76**, 155–160 (2004)

37. J. Yu, G. Dai, B. Huang, Fabrication and Characterization of visible-light-driven plasmonic photocatalyst Ag/AgCl/TiO_2 nanotube arrays. J. Phys. Chem. C **113**, 16394–16401 (2009)

38. P. Wang et al., Ag@AgCl: a highly efficient and stable photocatalyst active under visible light. Angew. Chem. Int. Ed. **47**, 7931–7933 (2008)

39. J. Guo et al., Highly stable and efficient Ag/AgCl@TiO_2 photocatalyst: preparation, characterization, and application in the treatment of aqueous hazardous pollutants. J. Hazard. Mater. **211–212**, 77–82 (2012)

40. W. Liao et al., Photoelectrocatalytic degradation of microcystin-LR using Ag/AgCl/TiO_2 nanotube arrays electrode under visible light irradiation. Chem. Eng. J. **231**, 455–463 (2013)

41. I. Liu et al., The photocatalytic decomposition of microcystin-LR using selected titanium dioxide materials. Chemosphere **76**, 549–553 (2009)

42. M.E. Van Apeldoorn et al., Review toxins of cyanobacteria. Mol. Nutr. Food Res. **51**, 7–60 (2007)

43. R. Zanella et al., Alternative methods for the preparation of gold nanoparticles supported on TiO_2. J. Phys. Chem. B **106**, 7634–7642 (2002)

44. W. Choi, A. Termin, M.R. Hoffmann, The role of metal ion dopants in quantum-sized TiO_2: correlation between photoreactivity and charge carrier recombination dynamics. J. Phys. Chem. **98**, 13669–13679 (1994)

45. V.S. Mohite et al., Photoelectrocatalytic degradation of benzoic acid using Au doped TiO_2 thin films. J. Photochem. Photobiol., B **142**, 204–211 (2015)

46. G. Wang et al., Integration of membrane filtration and photoelectrocatalysis using a TiO_2/carbon/Al_2O_3 membrane for enhanced water treatment. J. Hazard. Mater. **299**, 27–34 (2015)

47. D. Liu et al., Photoelectrocatalytic degradation of methylene blue using F doped TiO_2 photoelectrode under visible light irradiation. Chemosphere **185**, 574–581 (2017)

48. D. Liu et al., Enhanced visible light photoelectrocatalytic degradation of organic contaminants by F and Sn co-doped TiO_2 photoelectrode. Chem. Eng. J. **344**, 332–341 (2018)

49. M. Pelaez et al., A review on the visible light active titanium dioxide photocatalysts for environmental applications. Appl. Catal. B **125**, 331–349 (2012)

50. W. Yu et al., Enhanced visible light photocatalytic degradation of methylene blue. Appl. Surf. Sci. **319**, 107–112 (2014)

51. X. Yu, Y. Zhang, X. Cheng, Preparation and photoelectrochemical performance of expanded graphite/TiO_2 composite. Electrochim. Acta **137**, 668–675 (2014)

52. J. Qu, X.U. Zhao, Design of BDD-TiO_2 hybrid electrode with P–N function for photoelectroatalytic degradation of organic contaminants. Environ. Sci. Technol. **42**, 4934–4939 (2008)

53. J. Li et al., Facile method for fabricating boron-doped TiO_2 nanotube array with enhanced photoelectrocatalytic properties. Ind. Eng. Chem. Res. **47**, 3804–3808 (2008)

54. H. Sun et al., Preparation and characterization of sulfur-doped TiO_2/Ti photoelectrodes and their photoelectrocatalytic performance. J. Hazard. Mater. **156**, 552–559 (2008)

55. T.C. Jagadale et al., N-doped TiO_2 nanoparticle based visible light photocatalyst by modified peroxide sol-gel method. J. Phys. Chem. C **112**, 14595–14602 (2008)

56. R. Asahi et al., Visible-light photocatalysis in nitrogen-doped titanium oxides. Science **293**, 269–272 (2001)
57. L. Han et al., Photoelectrocatalytic properties of nitrogen doped TiO_2/Ti photoelectrode prepared by plasma based ion implantation under visible light. J. Hazard. Mater. **175**, 524–531 (2010)
58. L. Sun et al., N-doped TiO_2 nanotube array photoelectrode for visible-light-induced photoelectrochemical and photoelectrocatalytic activities. Electrochim. Acta **108**, 525–531 (2013)
59. N.R. Khalid et al., Carbonaceous-TiO_2 nanomaterials for photocatalytic degradation of pollutants: a review. Ceram. Int. **43**, 14552–14571 (2017)
60. K. Nagaveni et al., Synthesis and structure of nanocrystalline TiO_2 with lower band gap showing high photocatalytic activity. Langmuir **20**, 2900–2907 (2004)
61. S.U.M. Khan, M. Al-shahry, Efficient photochemical water splitting by a chemically modified n-TiO_2. Sceince **297**, 2243–2246 (2002)
62. L. Chen et al., Enhanced visible light-induced photoelectrocatalytic degradation of phenol by carbon nanotube-doped TiO_2 electrodes. Electrochim. Acta **54**, 3884–3891 (2009)
63. S. Ameen et al., Sulfamic acid-doped polyaniline nanofibers thin film-based counter electrode: application in dye-sensitized solar cells. J. Phys. Chem. **114**, 4760–4764 (2010)
64. W. Cui et al., Applied catalysis B: environmental polyaniline hybridization promotes photo-electro-catalytic removal of organic contaminants over 3D network structure of rGH-PANI/TiO_2 hydrogel. Appl. Catal. B **232**, 232–245 (2018)
65. X. Li et al., Efficient visible light-induced photoelectrocatalytic degradation of rhodamine B by polyaniline-sensitized TiO_2 nanotube arrays. J. Nanopart. Res. **13**, 6813–6820 (2011)
66. K. Yang et al., Materials science in semiconductor processing enhanced photoelectrocatalytic activity of Cr-doped TiO_2 nanotubes modified with polyaniline. Mater. Sci. Semicond. Process. **27**, 777–784 (2014)
67. H. Zhang et al., Dramatic visible photocatalytic degradation performances due to synergetic effect of TiO_2 with PANI. Environ. Sci. Technol. **42**, 3803–3807 (2008)
68. Y. Zhang, S. Geißen, C. Gal, Carbamazepine and diclofenac: removal in wastewater treatment plants and occurrence in water bodies. Chemosphere **73**(8), 1151–1161 (2008)
69. S. Yang, Y. Liu, C. Sun, Preparation of anatase TiO_2/Ti nanotube-like electrodes and their high photoelectrocatalytic activity for the degradation of PCP in aqueous solution. Appl. Catal. A **301**, 284–291 (2006)
70. A. Damin et al., Structural, electronic, and vibrational properties of the Ti-O-Ti quantum wires in the titanosilicate ETS-10. J. Phys. Chem. B **108**, 1328–1336 (2004)
71. Z. Zhang et al., Photoelectrocatalytic activity of highly ordered TiO_2 nanotube arrays electrode for azo dye degradation. Environ. Sci. Technol. **41**, 6259–6263 (2007)
72. B.G.K. Mor et al., Transparent highly ordered TiO_2 nanotube arrays via anodization of titanium thin films. Adv. Func. Mater. **15**, 1291–1296 (2005)
73. Y. Zhu et al., Sonochemical synthesis of titania whiskers and nanotubes. Chem. Commun. **2001**, 2616–2617 (2001)
74. X. Nie et al., Synthesis and characterization of TiO_2 nanotube photoanode and its application in photoelectrocatalytic degradation of model environmental pharmaceuticals. J. Chem. Technol. Biotechnol. **88**, 1488–1497 (2012)
75. J.J. Lopez-Penalver, Photodegradation of tetracyclines in aqueous solution by using UV and UV/H_2O_2 oxidation processes. J. Chem. Technol. Biotechnol. **85**, 1325–1333 (2010)
76. Q. Zhang et al., Applied surface science electrochemical assisted photocatalytic degradation of salicylic acid with highly ordered TiO_2 nanotube electrodes. Appl. Surf. Sci. **308**, 161–169 (2014)
77. W. Liao et al., Electrochemically self-doped TiO_2 nanotube arrays for efficient visible light photoelectrocatalytic degradation of contaminants. Electrochim. Acta **136**, 310–317 (2014)
78. M. Xing, An economic method to prepare vacuum activated photocatalysts with high photo-activities and photosensitivities. Chem. Commun. **47**, 4947–4949 (2011)
79. G. Wang et al., Hydrogen-treated TiO_2 nanowire arrays for photoelectrochemical water splitting. Nano Lett. **11**, 3026–3033 (2011)

80. M. Ferna, C. Belver, C. Ada, Photocatalytic behaviour of Bi_2MO_6 polymetalates for rhodamine B degradation. Catal. Today **143**, 274–281 (2009)
81. X. Hu, C. Hu, J. Qu, Preparation and visible-light activity of silver vanadate for the degradation of pollutants. Mater. Res. Bull. **43**, 2986–2997 (2008)
82. X. Zhao et al., Photoelectrochemical degradation of anti-inflammatory pharmaceuticals at Bi_2MoO_6—boron-doped diamond hybrid electrode under visible light irradiation. Appl. Catal. B **91**, 539–545 (2009)
83. X. Zhang et al., Effect of Si doping on photoelectrocatalytic decomposition of phenol of $BiVO_4$ film under visible light. J. Hazard. Mater. **177**(1–3), 914–917 (2010)
84. T. Chen et al., One-step synthesis and visible-light-driven H_2 production from water splitting of Ag quantum dots/ g-C_3N_4 photocatalysts. J. Alloy. Compd. **686**, 628–634 (2016)
85. J. Sun et al., H_2O_2 assisted photoelectrocatalytic degradation of diclofenac sodium at g-C_3N_4/$BiVO_4$ photoanode under visible light irradiation. Chem. Eng. J. **332**, 312–320 (2018)
86. S. Yu et al., Preparation and photoelectrocatalytic properties of polyaniline-intercalated layered manganese oxide film. Catal. Commun. **11**, 1125–1128 (2010)
87. F. Vicent et al., Characterization and stability of doped SnO_2 anodes. J. Appl. Electrochem. **28**, 607–612 (1998)
88. E.O. Scott-emuakpor et al., Remediation of 2, 4-dichlorophenol contaminated water by visible light-enhanced WO_3 photoelectrocatalysis. Appl. Catal. B **123–124**, 433–439 (2012)
89. S. Ghasemian, S. Omanovic, Fabrication and characterization of photoelectrochemically-active Sb-doped Snx-W (100-x)%-oxide anodes: towards the removal of organic pollutants from wastewater. Appl. Surf. Sci. **416**, 318–328 (2017)
90. S. Ghasemian et al., Photoelectrocatalytic degradation of pharmaceutical carbamazepine using Sb-doped Sn 80%-W 20%-oxide electrodes. Sep. Purif. Technol. **188**, 52–59 (2017)
91. L. Giudice, A. Pollio, J. Garric, Ecotoxicological impact of pharmaceuticals found in treated wastewaters: study of carbamazepine, clofibric acid, and diclofenac. Ecotoxicol. Environ. Saf. **55**, 359–370 (2003)
92. X.Z. Li et al., Photoelectrocatalytic degradation of humic acid in aqueous solution using a Ti/TiO_2 mesh photoelectrode. Water Res. **36**, 2215–2224 (2002)
93. Y. Xin et al., Photoelectrocatalytic degradation of 4-nonylphenol in water with WO_3/TiO_2 nanotube array photoelectrodes. Chem. Eng. J. **242**, 162–169 (2014)
94. D. Vione et al., Phenol chlorination and photochlorination in the presence of chloride ions in homogeneous aqueous solution. Environment. Sci. Technol. **39**, 5066–5075 (2005)
95. N. Lu et al., Synthesis of molecular imprinted polymer modified TiO_2 nanotube array electrode and their photoelectrocatalytic activity. J. Solid State Chem. **181**, 2852–2858 (2008)

Index

© Springer Nature Switzerland AG 2020
O. M. Ama and S. S. Ray (eds.), *Nanostructured Metal-Oxide Electrode
Materials for Water Purification*, Engineering Materials,
https://doi.org/10.1007/978-3-030-43346-8

Printed in the United States
by Baker & Taylor Publisher Services